One-on-One Combat in the Arena

One-on-One Combat in the Arena

Le Combat Seul à Seul en Camps Clos

Marc de la Béraudière

translated by Chris Slee

One-on-One Combat in the Arena
Copyright ©2024 Chris Slee (translator)

ISBN: 978-0-6452538-3-2 (eBook)
ISBN: 978-0-6452538-2-5 (Print)

All Rights Reserved. No part of this publication may be reproduced, stored in a retrieval system, or transmitted, in any form or in any means – by electronic, mechanical, photocopying, recording or otherwise – without prior written permission from the copyright owner(s).

Le Combat Seul à Seul en Camps Clos, Marc de la Béraudière. The original text is from a facsimile of the 1608 edition. It is asserted this book is in the public domain.

Created at LongEdge Press, first edition.

Contents

Introduction xi
   Marc de la Béraudière xi
   Description of the Text xii
   In the Cause of Social Harmony xiv
   Translators Notes xv
   Acknowledgements xvi

One-on-One Combat in the Arena 3
   To the King 4
   Extract from the King's Privilege 6
   To the Reader 7

Part One 9
   First Chapter – If combat must be permitted and if it is lawful   9
   Chapter II – In what circumstances the King must allow
      combat in the arena to his subject 11
   Chapter III – The causes for which one must allow combat .. 12
   Chapter IIII – The quality of persons, when, and to whom
      the duel must be granted 12
   Chapter V – Of those who are exempt from duels 13
   Chapter VI – About Chevaliers of the Order of the King and
      Captains of Gendarmes 15
   Chapter VII – The form that combatants must observe 16
   Chapter VIII – That out of revenge one must not undertake
      combat in the arena 17
   Chapter IX – If bastards must be accepted for combat in the
      arena 18
   Chapter X – About challenges 19
   Chapter XI – About seconds 21
   Chapter XII – About selection of the field 21

Chapter XIII – About the construction of the field and what it must be 22
Chapter XIIII – Who must have the field built? 23
Chapter XV – Whoever touches the ropes of the list or the palisade must stand defeated 23
Chapter XVI – That the defender must enter the arena first with the weapons with which he wants to fight 25
Chapter XVII – That the parents must not attend duels in the arena which are ordered 26
Chapter XVIII – That one must not speak after the Chevaliers have entered the arena 26
Chapter XIX – About that which it is necessary the victors observe on the day of the duel 27
Chapter XX – Of the choice and selection of weapons ... 28
Chapter XXI – About the proper weapons for the duel ... 29
Chapter XXII – If the Prince can Interrupt the Duel 32
Chapter XXIII – If the two Chevaliers, being in the arena, can change the quarrel 32
Chapter XXIIII – The rights which belong to Chevaliers after victory in a duel 34
Chapter XXV – Of he who will have accepted the duel but is not found on the assigned day 36
Chapter XXVI – About he who has been defeated once in the arena and, after being called to a duel by another, stood victorious, if he has by this recovered his honour 37
Chapter XXVII – If one must be received to fight the quarrel of another in the arena 40
Chapter XXVIII – The ways of giving the lie 42
Chapter XXIX – Of the form and manner of giving the lie and how the Chevalier must conduct himself 43
Chapter XXX – About the ambush 44
Chapter XXXI – Baton strikes and slaps 46

Second Part   **49**

Chapter One – About challenges in the King's Court 49
Chapter II – Of the several types of challenges 51
Chapter III – Of those who send *cartels* of challenge by their servants 53
Chapter IIII – Of those who go to challenge the enemy of their friend in his house 54
Chapter V – Of the one who will call out his friend's enemy in a house other than his own 55

Chapter VI – Of those who present themselves to second their companion 56
Chapter VII – If one must call out his enemy at the head of a company 57
Chapter VIII – If the yeoman should call out the gentleman to duel 58
Chapter IX – Of close relatives who call out each other to duel 60
Chapter X – Of two combatants, if the one who retreats must be accused of cowardice 61
Chapter XI – If one of the two who have a quarrel convinces the other to draw a sword, even though he doesn't want to do it, and the first wounds the second, it is badly done 62
Chapter XII – Of two who are in a one-on-one duel and one seizes the sword of his companion then hits him, whether he has fought well 63
Chapter XIII – If someone pushes you rudely from playfulness, must you call him out? 64
Chapter XIIII – If a gentleman in service must be considered equal in combat to a gentleman of honour and from a good house 66
Chapter XV – If a sexagenarian gentleman must be exempt from the duel 67
Chapter XVI – Of him who speaks on behalf of his friend . 68
Chapter XVII – How this word "to feel" must be understood 69
Chapter XVIII – How the Chevalier must feel when he is offended 70
Chapter XIX – Of the refusal many make in their quarrels and of asking their enemy's pardon 71
Chapter XX – Of the pardon that some ask for the satisfaction of their offences 73
Chapter XXI – Of those who do not want to confess the cause and subject of their quarrel 74
Chapter XXII – On satisfaction 74
Chapter XXIII – For gentlemen who had a quarrel to agree, it is necessary to know the basis and the origin of their quarrel 77
Chapter XXIIII – If it is necessary to make the two gentlemen which have had quarrel embrace after they have ben reconciled 78

Chapter XXV – If two kings should fight individually for
    the their states 79

Third Part   81

Chapter One - Of the status of the arbiters and their quality   81
Chapter II - That it is required to know the name of the
    arbiters 83
Chapter III - Of those who effect a reconciliation for an
    accidental quarrel 83
Chapter IIII - That one must keep his promise between gen-
    tlemen 84
Chapter V - When a gentleman has given his word, whether
    he is bound to name he to whom he gave it 85
Chapter VI - That the assemblies must be defended by all
    the Provinces 86
Chapter VII - What forms should the governors of
    provinces follow in order to appease the quarrels
    arising in their government? 87
Chapter VIII - That an honourable chevalier hearing
    [someone] speaking badly of his friend must respond
    to it 88
Chapter IX - If a prisoner of war, having given his oath,
    must keep it 89
Chapter X - Of him who leaves home to fight a battle, and
    cannot be found on the appointed day 91
Chapter XI - In the conduct of an army, which is most nec-
    essary: the bold man or the wise man 93
Chapter XII - On the difference that must be made between
    gentlemen who claim to be of better house than another 95
Chapter XIII - Continuation of this chapter and where this
    name of gentleman came from 97
Chapter XIIII - Whether someone is doing himself wrong
    if, being outraged, he employs someone greater than
    himself. 99
Chapter XV - If he who is envious of another must be held
    as an enemy 99
Chapter XVI - How these terms "You do not know what
    you are doing" must be understood 101
Chapter XVII - Of those who tell themselves, "I am a good
    man and a gentleman of honour". How this must be
    understood. 102

Chapter XVIII - If calling another angry must be taken as an injury 104
Chapter XIX - How these words should be taken, "Take it as you will" 105
Chapter XX - Of the fear that the chevalier might have of his enemy 106
Chapter XXI - That the weapons that chevaliers today present are not reasonable or in use 108
Chapter XXII - If someone, having bought a horse, is recognised and vouched for by an army, if he must return it 109
Chapter XXIII - A Captain who loaned horses and weapons to the soldier, should they be paid for? 111
Chapter XXIIII - That he is not to look at his friend's letters 111
Chapter XXV - It is very honest for gentlemen to salute each other 112

Fourth Part   115

Chapter One - The civil and internal wars are partly the cause of the abundance of quarrels 115
Chapter II - That the surest means of avoiding quarrels is maintaining piety. Justice is the true foundation of maintaining concord and friendship. 117
Chapter III - Of Justice which is companion of Piety 119
Chapter IIII - That the Nobility must be nourished in every honest exercise and learn that it is from virtue that one lives happily 121
Chapter V - Of the virtues that are proper to the gentleman to make him perfect and well accomplished 123
Chapter VI - That the boldness and valour of a gentleman should not be esteemed if they are not accompanied by magnanimity 125
Chapter VII - That it is something that greatly weakens the boldness of a gentleman, who does not move from his house, and who does not seek the hazards of war . . . 127
Chapter VIII - Following in this Chapter, proving that laziness in a Gentleman is to be avoided 129
Chapter IX - That there is more honour for the gentleman, when he is offended, to draw his justice from it with honest satisfaction than to avenge himself 131
Chapter X - That shame and dishonour must prevent the gentleman from doing wrong 134

Chapter XI – Of the fear that accompanies shame 135
Chapter XII – Of the fear and false shame which the gentleman must avoid 136
Chapter XIII – That the Gentleman must keep from talking too much because from that comes the increase of quarrels, especially when one speaks inappropriately . 137
Chapter XIIII – That the gentleman who writes inappropriately is greatly to be condemned, because from this arises many quarrels 140
Chapter XV – That ingratitude is a vice which the noble gentleman must avoid because he who is ungrateful procures many enemies 141
Chapter XVI – That the Gentleman must not reproach his friend for the pleasure he has given him 143
Chapter XVII – That the poverty of the Gentleman should not cause him to be poorly behaved, so that he does not fall into disgrace 144
Chapter XVIII – That the Gentleman should not set his heart on riches, if not to use them and follow virtue according to his quality 146
Chapter XIX – That the gentleman should follow wastefulness 149
Chapter XX – Of Recklessness 151
Chapter XXI – That memory is excellent to the Gentleman who wishes to follow arms. And that there have been great Captains who have been much esteemed. 153
Chapter XXII – That the Gentleman in his affairs must seek the advice and counsel of others 155
Chapter XXIII – That the Gentleman should not be curious about other people's affairs 156
Chapter XXIIII – That the gentleman must guard the honour of the Ladies and fight for his mistress 157

LongEdge Press   **161**

Introduction

I have a habit of saying that each text I translate is remarkable. The freedom of the dilatante allows me the choice to translate only the books that interest me and are relavant to my historical fencing practice. This book, however, is indeed remarkable for the insights it provides in to the society, culture and politics of late 16th century and 17th century France, and world of the French court.

Marc de la Béraudière

The titles Marc de la Béraudière held reflected his standing in society. He was the lord of Millac and Mauvoisin, and held the rank of Chevalier of the Order of the King, his command extending to fifty men-at-arms. De la Béraudière married Renée du Chiron, and produced a daughter named Jeanne, who would go on to marry Philibert de Maroy, from d'Aunis, holding the titles of Lord of La Grange and St Vivien. Their lineage would continue through their son, Emery-Philibert de Maroy, who inherited the lordship of Millac and other prestigious domains. Historical records, preserved in the Notary Archives of M. Chaigneau, highlight Emery-Philibert's connection to that chateau.[1]

De la Béraudière served as a Gentleman Ordinary of the Chamber, showing his close proximity to the King's inner circle. His château in Isle-Jourdain in the Basse Marche demonstrates his prominence in the region. De la Béraudière sold the land and lordship of Mauvoisin in the Duchy of Chatellerault to Robert de Combault in 1580, a fellow knight of the Order of the King and advisor in the King's private council. The reasons for the sale are unclear but may be part of the

---

[1] "Millac Et Son Passe." *Millac: Site Officiel*, www.millac.fr/ma-commune/pages/850-millac-et-son-passe. Accessed 22 Jul. 2023.

dowry of Louise de la Berauldière.[2]

Description of the Text

One of One Combat in the Arena fits into the broad category of "advice literature" but beyond this simple definition, exactly where it fits in the established literary categories is unclear. On the one hand, it fits the "mirror for princes" category by providing clear instructions to assist the prince in properly governing his realm in the areas of justice, keeping the peace, resolving quarrels, noble precedence in quarrels, and administering just punishments for infractions. There are strict procedures to be followed when the honour of one party is injured by another. The aim is to prevent civil disorder and to reinforce the authority of the prince as the *pater familias* of the body politic. On the other hand, it is also a book which conforms to the courtier's book of etiquette genre in that it outlines very clearly how a gentlemen of the court should act in the presence of both the prince and other nobility. It discusses the concepts of honour, duty, justice and mercy to provide the courtier clear guidelines for proper conduct in word and deed and the remedies which may be reasonably sought when this decorum is breached.

The book is both a reaction to and an attempt to localise Muzio's *Il Duello* to the experience of the French upper classes. Girolamo Muzio, known also as Mutio Justinopolitano, was an Italian courtier, poet, and author whose contributions played a pivotal role in the defence of vernacular Italian against Latin during the Renaissance. Among his most renowned publications were *Il duello* (1550) and *Il gentilhuomo* (1571), which both delved into themes of honour, nobility, and chivalry. He was intolerant towards those who opposed his views, particularly heretics or individuals who dared to challenge his writings on chivalry. Italian etiquette and social practice of fifty years earlier are not those of contemporary France. Scattered throughout the text are passages where Béraudière says "I agree with Muzio on this point when he says …." and "I disagree with Muzio here for these reasons." These passages are useful in highlighting the aspects of social thinking and practice which differ from those of Italy. These passages are worthy of deeper study in order to develop a better understanding of

---

[2]"Châtelet de Paris. Y//119-Y//123. Insinuations (19 Septembre 1577 - 15 Juin 1582)." *France Archives: Portail Nationale Des Archives*, francearchives.gouv.fr/fr/facomponent/4b7c84bd35887a54b1615f6ef1af5fcd5340404b. Accessed 22 Jul. 2023.

the French context.

Béraudière's text is divided, rather arbitrarily, into four sections. The themes of the work do not map easily across the four sections with some themes spanning two sections and multiple themes contained in a single section. This makes the divisions between the sections more likely to be about the mechanics of printing the book rather than any design principle.

The first section of the book focuses on the regulations and laws governing combat. It explores the legality and circumstances under which duelling is permitted, examining the causes and qualities of individuals for whom the duel must be granted. Additionally, it explores the exceptions and conditions under which individuals are exempt from participating in duels. This section also outlines the role of participants' social status, particularly Chevaliers of the Order of the King and Captains of Gendarmes, shedding light on the significance of certain social or military statuses in the context of duelling. Furthermore, the chapters dedicated to duelling procedures and etiquette elucidate various aspects such as the formalities observed by combatants, the selection and preparation of the duelling field, weapon choice, rules for entering the lists, and expectations for the day of the duel. The rights of the victors and the consequences of certain actions during the duel are also discussed, emphasizing the importance of honour and reputation.

The second part of the book explores various types of challenges issued in different locations and contexts. Social dynamics, power structures, and familial relationships within duelling are examined, along with ethical considerations and questions of fair play. The section delves into exemptions, age related considerations, and the equality of combatants. It also addresses topics such as offense, feelings, and the interpretation of words. Moreover, protocols for refusing quarrels, seeking pardon, and finding satisfaction are discussed, explaining the process of reconciliation and the broader context of duelling, including its connection to legal appeals and hypothetical state conflicts. This section provides invaluable insights into the ethical considerations surrounding the duel.

The third section of the book explores the status and qualities of arbiters, highlighting the importance of knowing their names. It explores the process of reconciliation in accidental quarrels, emphasizing the significance of keeping promises among gentlemen and examining the obligations and responsibilities tied to one's word. Furthermore, the role of governors in resolving disputes within their provinces is discussed, along with the duty to defend assemblies. The section also touches upon honorable responses to someone speaking ill of a friend, the obligations of a prisoner of war to honour their oath, and matters related to military conduct and accountability. Distinctions between individuals claiming superiority in lineage are examined, as are the consequences of seeking revenge through someone of higher status. Envy and its implications in relationships are considered, providing insights into period conceptions of the human psyche within the context of duelling. Additionally, the section explores the interpretation of phrases, anger related insults, and the significance of fear for knights and combatants. It also covers obligations regarding borrowed horses and weapons, respect for privacy, and the importance of salutations between gentlemen.

The final section of the book looks at the relationship between civil wars and the abundance of quarrels, stressing the importance of maintaining piety and justice as means to avoid conflicts. It explores the connection between justice and piety, and the nurturing of nobility through virtuous exercises. The section highlights virtues that contribute to the perfection of a gentleman, emphasizing the importance of magnanimity, boldness, and valour. It discusses the negative impact of idleness and laziness and the need for a gentleman to avoid such vices. Seeking satisfaction through honest means rather than seeking revenge is encouraged, and the role of shame and dishonour in preventing wrongdoing is examined. The section also condemns excessive talking, inappropriate speech, and ingratitude, emphasizing the importance of maintaining good behaviour despite poverty and using wealth virtuously. The concept of recklessness is explored, and the importance of memory, seeking advice, and defending honour is discussed. In summary, this section addresses personal conduct, virtue, honour, and relationships, offering guidance to gentlemen on navigating societal challenges and maintaining harmonious interactions.

In the Cause of Social Harmony

At the level of the state, de la Béraudière discusses the relationship between civil and internal wars and the abundance of duels. De la Béraudière argues that civil and internal wars contribute to the prevalence of duels due to the diversity of opinions among the involved parties and the resulting envy and conflicts. While he never actually mentions the conflicts and atrocities that modern historians know as the Wars of Religion in the France of the period, it is obvious that he is referring to these scars on the nationale psyche. De la Béraudière points out that civil wars have historically been the downfall of kingdoms, monarchies, and republics. Examples given are Athens, Sparta, and Rome, where despite their achievements and successful governance, they eventually succumbed to internal strife and division. De la Béraudière mentions the rivalry between the Guelphs and Ghibellines in Rome and the prolonged conflict between the Houses of Burgundy and Orleans in France, resulting in extensive bloodshed and turmoil. He concludes by emphasizing that civil wars have given rise to the current state of duels and argues that a wise legislator needs to be called upon to find a remedy, and restore peace and harmony within the kingdom.

On a personal level, de la Béraudière argues that maintaining piety and obeying justice are the surest means of avoiding quarrels and duels. Piety is defined as possessing virtues such as being good-natured, charitable, peaceful, and temperate. Individuals who are merciful and exhibit piety are inclined towards friendship and peace, avoiding quarrels and offenses. The company of virtuous individuals is desired, while quarrelsome and foolish individuals should be avoided. Good company encourages goodness and honour, while bad company can lead to dishonour and quarrels. Humility is emphasized as a virtue that complements piety, allowing a person to maintain good relations and be respected. Arrogance and glory seeking lead to being disliked and neglected, causing quarrels and inconveniences. Concord and union among relatives, friends, and neighbours are essential to avoid divisions and divorces. De la Béraudière recognises the challenges of human life, but suggests that wisdom, magnanimity, and virtue can resist the changing nature of circumstances. The preservation of relationships with relatives, neighbours, and friends is achieved through piety, divinely guided and upheld with close friendship.

Translators Notes

Béraudière's book is easy to read although, as always, punctuation in the text is idiosyncratic and more related to pauses when reading aloud than to the signalling of completed ideas. At times, the reader needs to exert effort to read around a chapter in order to uncover the links Béraudière makes between the subjects of sentences and the topic under discussion. It is sometimes simply not clear how one sentence relates to the next, and only having completed the whole paragraph does the meaning become clear.

Acknowledgements

This book owes its clarity and cohesiveness to the invaluable contributions of Lois Spangler, an exceptional editor whose keen insights and meticulous work breathed life into these pages.

One-on-One Combat in the Arena

One-on-One Combat in the Arena

By My Lord MARC de la BÉRAUDIÈRE, Chevalier of the Order of the King, Captain of Fifty Men-at-arms of his Ordonnaces, Lord of Mauvoisin, with several questions proper to this subject, together [with] the Gentleman's way of avoiding quarrels, to leave them with his honour.

Divided into four parts

At Paris, at the house of ABEL L'ANGELIER, at the first pillar of the Grand Salle of the Palace.

MDCVIII

With the King's Privilege

To the King

Sire,

since this miserable custom has taken so privileged a path, regardless of any prohibition that Your Majesty can make, [and] the Chevaliers cannot reform themselves from entering into proofs of arms, it is reasonable that you actively embrace this cause, performing the office of the Sovereign, in order to conclusively judge to whom justice[3] belongs. You could not do anything more divine, Sire, than to recognise the quarrels and duels which are made day on day in your Kingdom. The number [of them] and the disorder which they cause are a shame to the French nation and a horror to all other nations who hear talk of the single combats[4] which are habitually exercised here, and most often with very little foundation. I speak of this with the desire I have that Chevaliers can take again the route of those who have acquired renown as valorous knights. I cannot invent any means nor find a way which is more useful other than it please Your Majesty to command that no Chevalier should ever be called upon to enter combat without your permission, on pain of his life. Then, you will see quarrels abate little by little and the Nobility take up an exercise which is much more honourable than killing each other. Today, we see that for simple, very light words, and perhaps carrying no harm to one or the other, they are thus immediately to hands.[5] And [they] have this opinion that no Gentleman can be held or be esteemed valiant if he has not tested his courage with someone. This is a strange opinion and which in truth must be suppressed by your authority. I know that the status of Chevalier is a rank of honour, belonging only to Chevaliers who enter tests of arms. Also, it's necessary that they match their courage with others who only have honour by recommendation. Let them be honourably grounded in their quarrels and have your permission. And all the more, Sire, as you have the power to give such laws as you please about this disorder, I have not been afraid, entirely confident in your great goodness, of dedicating this present tract to you. Although it may not be accompanied by great knowledge or beautiful language, it is solid in my hands, just like a simple soldier. I have made it for the love I have for arms and for the good and honour of all generous men. You would do me the honour, if you please, Sire, of having it and receiving it from he who wishes to remain until the

---
[3] *droict*
[4] *combats particuliers*
[5] that is, drawing weapons

last breath of his life

Your very humble and very obedient subject and servant,
Marc de la Béraudière, Lord of Mauvoisin.

Extract from the King's Privilege

By the King's Grace and Privilege, he permits Abel l'Angelier, master bookseller[6] in the University of Paris, to print or have printed the present book entitled *One-on-One Combat in the Arena* by the Lord of Mauvoisin, and has made express prohibition to all booksellers and printers to neither print nor sell nor distribute the said book without the leave and consent of the said l'Angelier, and this for the term of six years on pain of arbitrary fine and of the confiscation of all the books which are found. And finally we want that this present extract of Privilege being declared in this book be duly signified to all booksellers and printers of the Kingdom, as is more fully declared in the King's Letters Patent given at Paris, 17 April 1608.

By the Council.

Signed *Brigard*

---

[6] *Libraire iuré*

To the Reader

Learn arms, reader, to protect life and reputation.
You strengthen everyone with the protection and art of Mars.
Justice [is] yet always a struggle. Take up arms
In just war. The gods help no one.
In this way, you may hold a career so distinguished, blessed,
Showing itself the leader and partner of the chosen life.
Make a memory of reading this book. You may store [it] in the mind.
From this, believe me, greater glory will be yours.

— F. de la BERAUDIERE, Abbé of Nouaillé Abbey.[7]

---

[7] A defunct Benedictine abbey known as *Abbaye de Nouaillé* and *Abbaye Saint-Junien de Nouaillé* in the Diocese of Poitiers (*Nouaillé-Maupertuis, Vienne*).

Part One

First Chapter – If combat must be permitted and if it is lawful

Duels have been condemned by all divine and human laws as something which is contrary to the law, principally, the law of the Church and the commands of God, that to call out and kill each other, as he [does] who desires to spill the blood of another, is to be willingly subject to die by the sword. Our Lord does not want the shedding of blood, nor that man demand justice[8] against the life of his fellow among Christians who must live in common accord and maintain themselves in peace and union and live according to the commandments of God. The Persians, Hebrews, Greeks and Latins very expressly forbade it, if it were not in a legitimate, good and well-founded war and [done] in order to end it, like the duel between David and Goliath in Kings I, chapter 17 and the single combat[9] between Hector and Ajax reported by Homer in the tenth [book] of the Iliad, and like the duel of the Horatii against the Curiatti told by Titus Livy, like the Fabians who battled Curiatti before the assembled battalions[10] in order to end the dispute between their countries and nations, [like] Romulus fighting Titus, King of the Sabines. Similarly, even a King conducting his army, and being there in person, fights another King against whom he was at war, in order to end by arms, both alone, the quarrel and dispute which they have and avoid a greater waste and loss of their men to come. And if the one-on-one combat was done otherwise, it closely imitates bestial brutes who with ferocity are held and are hit because they have neither the reason nor the judgement to discern the evil and the shame which comes from it. Thus the reason

---

[8] *intenter*
[9] *monomachie*
[10] *les batailles rangees*

why the ancient and very civilised nations condemned and detested all individual combats. However, fighting in the arena has always been a common practice among the French, English, Burgundians, Italians, Germans, and Northerners. It has been received, observed, and undertaken with many good considerations and great, evident reasons. Seeing the country filled with brave gentlemen and good soldiers, who are well drilled and instructed in arms and continually exercised in making war, this exercise and practice of arms is considered honest. By imitating this practice, they will better make and acquire the most honourable reputation, the title of an honourable and valiant man. Therefore, they are taught to know and debate virtue and honour. In this way, the Gentleman and brave soldier, well tested, jealous of his honour and of his merit and valour, will allow nothing to be said by his companion which he thought could offend his honour and reputation. It is the circumstance of the quarrels which arise between them that leads them to call each other to combat when they understand that their honour is offended. On this, many murders have been committed. Thus why the King must be careful, in order to prevent this insolence, to have an eye that his subjects cannot be called out without his permission. Otherwise, there would be confusion and almost brigandage in his Kingdom if such a way of being called out to fight was not suppressed by the Prince. This makes me of the opinion that it may be better to grant the duel to his subjects who are of the condition and the profession of honour than to endure such miseries and misfortunes which happen in his Kingdom. In denying the duel, a King of the Lombards named Rotaris wanted to remove it from his subjects. But he was forced to undertake it again even though he protested that it was against all humanity. Philippe le Bel[11] forbade them in his Kingdom. But his subjects immediately begged him to bring them back in order to avoid the murders which took place every day. King Francis I, being a virtuous and very Christian Prince, permitted them several times in his Kingdom and, in his time, the Prince of Melphe, his Lieutenant in Piedmont, in order to suppress the insolent, who were usually there and cut the way to quarrels that occurred there, ordered a place where soldiers fought with the express restriction of only undertaking [this] with his permission. Then King Henry II permitted the duel at the beginning of his reign and later he forbade it by an edict [continued by] King Charles IX, his son. This prohibition has been the cause of many murders which were done since, which are made and will be made if there is not otherwise put

---

[11] Philip IV of France (1268-1314)

in place management and such order that one can have reparations for injury to his honour — which seems to me can be done if the King alone permits one-on-one combat in his Kingdom in the arena with very rigorous prohibitions on not being called out otherwise, and that whoever will be called out without his commandment will be exemplary punished by his justice.

Chapter II – In what circumstances the King must allow combat in the arena to his subject

When I say that the King must allow combat in the arena to his subjects, I do not mean that it be allowed to all who would ask for it but, when asked for combat, the King must examine the cause of the quarrel and try by all good ways to reconcile[12] them and to call the marshals and principal Councillors in order to give them an accord to be kept without favour or affection, particularly [to award] the right to whom it belongs and to condemn the one who is wrong. And, if the Prince knows that it will be too difficult for the wrongdoer to embrace the right of equity, not wanting to submit to the judgement of his good council for whatever good reasons and remonstrates, then one could say, "the King must use his absolute authority and force him to temper this with the reason, that is to say, to hold the strong hand in order that right and reason be maintained." But I say that coming to this, he must seek all means of bringing them to accord. And if the quarrel is of such consequence and so difficult that he who demands the duel is so insulted and offended, and that he goes there with his honour only able to be satisfied by arms, immediately beseeching the King to permit him the duel, it may be difficult for the latter,[13] wanting to support the honour of his subject and also [when] by refusing,[14] dishonouring him, because in truth the King has a lot of power over his subject although he goes with his honour, it is necessary that the subject debates it with his sword. For this consideration, it seems to me that [the King] must permit combat in the arena in order that his honour is restored.

---

[12] *appointer*
[13] *luy dernier*
[14] *veu qu'il y va de l'honneur de son subiect et aussi en luy refusant*

Chapter III – The causes for which one must allow combat

It is therefore required that the King look to the causes which are permitted to allow combat: the accusation of a crime of *leze majesté*[15] is legitimate for allowing the duel; also being accused of having committed a murder by ambush; of having wanted to commit treason, either on the person of the King or of having wanted to take money in return for a position, or having taken and stolen the King's money, [or] when one has defamed and dishonoured a lady. In all these cases, if one is accused, the King can permit the duel in order to defend the contrary. However, it would only be reasonable for the King, on a simple accusation, to order the duel with moderation. But he must proceed there, if on reflection and with such truth that the accuser is obliged to support his words with arms, in the case that he can find no witnesses who can testify with full proof to his accusation. For if the accusation laid against him deserves death, the duel must be granted.

Chapter IIII – The quality of persons, when, and to whom the duel must be granted

The granting of the duel is a form of justice that the King must observe for the preservation of the honour of his subjects, as he is the true and sole judge of his subject's honour. For, in truth, before the King orders a duel, many things must be considered; namely, if the combatants are of the same grade, if the duel that one demands is just, and if it must be granted and suffered that they come to arms. This was the reason that Philip, Duke of Burgundy, issue of the House of France,[16] forbade and abolished all duels in Holland in case a yeoman,[17] at any time and for little reason, called a Gentleman to duel. Also, it is not reasonable that a yeoman or someone from a very low place and without experience call out another who is a man of honour, of merit and of valour, and who has proved his person through many long years, being dignified with great merit. Such people have respect, and the King must have regard to their status. And if it happens that one who has less and lower status than he who is called to combat and, if he has no rank or experience because of his tender youth which rubs and

---

[15] A crime against the dignity of the Crown. In other words, treason.
[16] Possibly Philip III, the Good, b.1396, d.1467. Duke of Burgundy from 1419
[17] *le roturier*

provokes immediately[18] and takes quickly a quarrel in a good or bad cause, being confident in his skill and his valour or disdaining him with whom he has a quarrel and laughing at him, this must be corrected by the Prince. And when such a dispute is created between two Gentlemen, who are not equal in status or experience, the house and nation of he who has the greatest blame is obliged to satisfy it without entering into a duel and make him content. But, someone could object to me that this would be the way to give the reins to rich Gentlemen, of shouting down[19] one who has not similar status and quality. The response is that if the rich Gentleman or someone who has more honour than another has so forgotten injuring one lesser than him, or that with gaiety of heart he [the rich Gentleman] imposed on him some injurious and defamatory words that he [the injured party] has never forgotten he said, in this case, he [the rich Gentleman] is required to defend himself and deny the words which he put to him [the inured party], together fending off the injury which he [the rich Gentleman] did, offer to prove the contrary to him, and avenge himself by arms in the arena with permission of the King, providing that the accusation was worthy of death. But otherwise he who is not of similar status is held to respect and honour one who is of more than him. One must not accept the excuses of a heap of scroungers and mockers[20] who are used to jeering and after they have offended an honest man and honourable Gentleman, they think are acquitted of it by saying that they did it either in jest or without having thought about it. Thus the good excuses they give in such quarrels. When one reconciles them, they should examine their language before speaking, for once a word is said it can no longer be revoked. It's necessary keep it between Gentlemen for he who denies that which is said does a very great wrong to his reputation. However, I will say that if the word which he said is not true, he will have more honour in disclaiming it than in maintaining it. One will never have honour in maintaining a wrong cause. Also, be fully assured that God will never favour him.

Chapter V – Of those who are exempt from duels

It is very reasonable to observe the nation, status and house of those who claim to ask for the duel. For if it was allowed to all persons

---

[18] *qui le grade et chatouille*
[19] *de braver*
[20] *un tas de gaudisseurs et brocardeurs*

to be called out without exception, this would make for confusion. Princes are exempt from duelling in as much as it is appropriate for Gentlemen to quarrel with a Prince rather than respecting and honouring him as being of a greater and better house. When I speak of Princes, I intend those who are Princes of the Blood of France and those who are Princes of name and of arms and come from a previous ruling house. Truly to those, Gentlemen do not become equal nor are they called companions. And however much that a Gentleman has married a Princess or that the King has made a Duke and given much pre-eminence, he is not made equal to a Prince because of that. And if a difference occurs between the Prince and the Gentleman, the King must maintain a strong hand that the Gentleman respect [the Prince] and give him his due.[21] Officers of the Crown of France are similarly exempt from duelling. Besides being Gentlemen, they are honoured with greater status by the Crown. This was certified by King Charles VI for a quarrel between the Constable of Clifton and Jean de Craon, who resented with some displeasure that which the Constable had done to him, finding it appropriate, gave him strong sword blows. The King, not liking such an act, became so angry that he became frenzied. This is more amply written about in *The Chronicle of France*. I saw in my tender youth the Lord Chambray, Gentleman of Normandy, pursue a duel with our Admiral, Sir d'Annebaut. It was during the reign of King Francis I. The King rejected it and said that if there was a Gentleman in his Kingdom who had the temerity to pursue and demand single combat with Officers of the Crown, he would lose his head. But if it happened that an Officer of the Crown was offended by a Gentleman of honour, making it heard, he [the King] would decide the right as the facts merit and without coming to arms. Thus, the word and judgement of this generous Prince. I say further that the Officers of the Crown cannot and must not call a Prince to combat. And even though they are put and seated in the highest level and offices by the Crown, he [the Prince] must respect them. The reason is that when Princes die they always leave their children Princes. He or those who hold the Estates of the Crown possess them only as long as their lives and their children's [lives] last, afterwards rarely possessing them if not for the liberality of the King, to some and not to all.

---

[21] *le rendu bien comptant*

Chapter VI – About Chevaliers of the Order of the King and Captains of Gendarmes

Chevaliers of the Order of the King, being appointed and qualified by reason of their good and great services, must be exempt from duelling and respected in that they carry the Order of the King, their lord, which is the mark by which he wants to honour his good and loyal servants, together with those of his privy council, whom he chose and had elected from the number of men of honour, wise and well advised and proven, in order to be put near his person to decide all business which relates to the good of the Kingdom and his service. And regarding Captains of Gendarmes, it is not appropriate for the Gentleman who is without status to call them to duel. For any Gentleman who commands in general cannot be called out by one who is lesser than him and no matter whether he was[22] a Lieutenant of Gendarmes, Ensign, Standard Bearer, Captain of Foot, Camp Master or other similar rank, he cannot call out a Captain of Gendarmes to the arena. And the King must not allow it and must maintain them in order that his servants, thus appointed, are better recognised than they [the Gentlemen without status] are. It is very reasonable to honour those who are raised in status by their valour and great merit. We must not disdain those who the King has called close to his person in order to serve him. I remember on this topic a quarrel in Piedmont between M. de Vassay and Captain Moumas. The said Moumas asked the King [for permission for] a duel in order to have the right from M. de Vassay for the injury which he did him. But, by the advice of the Constable, Marshals and all the Council who were assembled several times for this instance, it was said that Moumas cannot and could not call out M. de Vassay to duel because he was not of his status or even of the same rank. Thus, the Captain of Gendarmes must be respected and honoured. But if a quarrel happens between two Gentlemen, one a Captain of Gendarmes and who must satisfy this Gentleman with whom he has quarrelled if it is with him to do so, it should be that he [the injured party] is held content and well satisfied, according to the advice of the King and his Council. Also, the Captains of One Hundred [and] Gentlemen of his house are not required to invite to combat one lesser than them, nor Captains of the Guard either. But the King can give [permission to] combat to those who are of the same rank. And in order that you recognise in what honour one held the Captains of Gendarmes, I put in this rank the captains who were greatly hon-

---

[22]*combien qu'il fuit* - suspect a spelling mistake and *fut* was intended

oured in my time. I saw M. de Sensac commanding the light cavalry. Princes of the Blood and great lords honoured him greatly and he was very well obeyed. It was during the journey of Valencienne, during the reign of Henry II, in the absence of M. d'Aumasle, who was a prisoner in the hands on the Marquis of Brandenburg, I saw M. de Desse, Lieutenant for the King in Scotland, who then was only Captain of Gendarmes, commanding a good army and many great lords. Also, the Marshal of Termes and he, both, were at the time only Captains of Gendarmes and on their return were made Chevaliers of the Order of the King.

You can see by this that the Order of the King has always been honoured and that after great and long service these Kings wanted to honour and recompense the services that these good and loyal servants have done by giving them their Order.

M. de Sensac also on his return from La Mirande had the Order with fifty men-at-arms. M. de Morluc also had the Order from the King when he returned from the Siege of Sienna with fifty men-at-arms. The King, therefore, must greatly esteem them and preserve them in their honour, and then they are called by their merits and are not allowed to be called out to duel as if they were simple soldiers. It is the reason why I wanted to write about it so that if a quarrel occurs between two who are not of the same rank, that they may be excluded from permission to duel in the arena. For I hold that all duels must be between equals and that the status of Chevaliers must be observed. All the more so if it is a rank of honour which has been acquired by a brave reputation among valiant men and persons of honour, who with arms have proven their valour and for this reason they have been put into the rank of Chevaliers. It is [also] the reason why the infamy of traitors against the King, thieves, all those who have been rejected from participation and use in war and stripped of arms must be exempt from duels.

Chapter VII – The form that combatants must observe

After having spoken of the reasons why the King must allow duels to his subjects, it is necessary at this time to speak about how the parties should fight and the form which they must observe. The aggressor must propose to the Prince that he maintains about his adverse party that the crime, which he accuses him of, is true, and he does not have witnesses who can testify about his words but, nevertheless, he wants to expose his life in proving to his adversary, with arms, that which he has avowed about him. And [he] very humbly entreats the King that he do him the honour of granting him the duel and that he does not want to enter combat if the accusation he puts forth is not true, being assured that God will so favour him in this quarrel that he [the petitioner] will make known to his adversary and to all attending that his words contain truth. The defender will respond to the King that he is innocent of all that which has been put to him and that it is a thing wickedly invented, begging his Majesty to preserve for him his honour and that it please him [the King] to grant the duel and asking for a day and place in order to be found there, promising on his honour not to miss it on pain of being stripped of all honour and to never again bear arms.

Chapter VIII – That out of revenge one must not undertake combat in the arena

Vengeance is forbidden by God and He has reserved it for himself. He corrects and chastises the proud and puts them under His power, for thus all vengeance should be put in His hands and should not be undertaken without His support, only undertaking it to defend others and not to fight for vengeance but only for a good and just cause. Sometimes, the injuries are so great that the Chevalier of honour, for his reputation and for the duty of chivalry, must with legitimate reason take up arms to attack and defend the injury that someone has done him. This should not be called vengeance but repelling the injury or outrage which has been received. For example, if someone wanted to rape his neighbour's wife, it is reasonable for him [the Chevalier] to by outraged by it and that he repulse this injury or infamy that he [the malefactor] wanted to do, not to be avenged but for his honour and for all posterity. It is a generous soldier's act to ready himself to take up arms against those who attempt to bring him infamy and dishonour, and when he will be avenged against this injustice, invoking the total power of God, who always preserves the right to whomever it belongs, one should not doubt that He will assist him. I will produce another example of he who undertakes combat about things which do not concern him.[23] Someone has killed a man which no one knows but me. I must not accuse him in order to avenge myself but because it is an act displeasing to God and because murderers must be

---

[23] *de chose qui ne luy touche en rien*

chastised. I want to enter into a proof by combat with the one who committed such an act and affirm to him that he did the homicide. For duels were only created for the justification of the truth and not for vengeance. The King must not grant the duel to anyone who wants to fight for vengeance, otherwise it would make a butchery of the arena. Likewise, there are several other effects which could arise even from adultery or treachery as well as from all villainous acts. I will not go to so many pages of Scripture nor search out such beautiful and holy reasons which forbid adultery, homicide, theft, the ravishing of honest girls and modest women, and all these types of villainy. It is quite certain that such acts are infamous and punishable, which the valiant man must repulse with arms. I would freely ask that if someone kills my father treacherously,[24] how should I be avenged? I mean, if I have no witnesses to prove this is murder. In this case, I could dispute it civilly as well as by arms and could undertake [this], according to God and men, to draw out its cause. According to God, because he allows revenge to be had in so holy a quarrel. According to men, because one would hold me of little valour and esteem if I did not have reason for this homicide. Also, for brothers and other relatives who love and have affection for you and who know that you have killed treacherously. But if the murder had been done head-to-head,[25] without any advantage, I would debate that he should not have any resentment because the quarrel has been finished by arms and one-on-one, which is a legitimate duel.

Chapter IX – If bastards must be accepted for combat in the arena[26]

This is a question which must be put to the rank of the Chevaliers. It is also the reason that one should make distinctions between bastards: because some are yeomen and others are issued and born of Gentlemen. These latter should have more privileges of arms than the other, since the noble resentment must touch them more. And this fine notion[27] must serve them as an example for acquiring honour and moving themselves to virtue. However, the bastard by law is excluded from all paternal and maternal succession because he is not legitimate. Neither being thus legitimately born nor under the con-

---

[24] *proditoirement*, by analogy with *proditeur*, traitor?
[25] *cap à cap*
[26] Original text has this chapter mislabelled as X
[27] *belle imagination*

dition of marriage, there is no law or reason which allows that he can call a Chevalier to combat and proofs by arms. It is true that there are bastards of such valour and such brave testing, and who have left such a reputation in the pursuit of arms, that they must not be dismissed from the duel when it will be a question of debating a quarrel. There are laws and Doctors [of Law] who speak in favour of bastards, and others who are against them. And in order to make a conclusion on all their opinions, mine would be that the fathers tried to legitimise their bastards in order to render them capable of possessing that which they could bequeath to them which could not, however, interest the true children issuing from a legitimate marriage. And thus coming to be legitimised by the King, I believe and hold this opinion that they can enter into proofs of arms and into the arena. There are found enough bastards who were Kings, Princes and great lords and sovereigns – even who have contested Kingdoms and have been received and ruled them and their posterity happily. I conclude, therefore, that the acknowledged bastard[28] from a noble line and legitimised by the Prince can enter the arena in order to contest his honour, especially when[29] he has acquired the honour and honourable status which have pushed him to this high rank of honour and virtue, which must make them prizes and honoured by all the brave Chevaliers. Priests' bastards must be excluded. Wherein, I would be of this opinion that the bastard of one who was not married, nor the mother also, should rather be legitimised by the Prince and received to the ranks of honour and knighthood than should he who is conceived in adultery, especially as it is a vice which is forbidden by the law of marriage, which further aggravates the sin. And then he who is thus elevated by the Prince can also be called to the ranks of honour and honourable duties, and in consequence must be received to contest his honour like other Chevaliers. The duel is nothing more than proving his honour with weapons, which is only appropriate for Chevaliers. And knighthood is a rank of honour and it is only lawful to anyone to enter into proofs of arms who has acquired this honour.

Chapter X – About challenges

There is today a custom for entering into quarrels that the parties are summoned by challenges.[30] This way of doing is very good –

---

[28] *le bastard aduoüe*
[29] *mesmement quand*
[30] *cartels*

which Chevaliers must observe in their quarrels. For, like in a civil case, one proceeds to action by summonsing the parties to respond to it. So, in duelling, which we all hold to be a form of justice, one should call out his enemy by a challenge in which should be put an explanation of their quarrel, as briefly as possible and in intelligible and common terms, without omitting anything in order that after being made none may gainsay it, that the assailant will call out his enemy by challenge to the duel, after having had the permission of the King. And the defender will respond to be there and maintain the contrary of all that with which he is charged[31] and can add a denial without going further, and, when it is put to him, he says that he will support the denial with weapons in his hands. He has not lost the choice and option of weapons, which stands always with the defender, since he is responding to the challenge of his adversary who calls him to combat. He accepts it when he responds to [the challenger] that he will be there with weapons. By this, he demonstrates his will and his valour.

The lord Hierosme Muzio in his book on Combat, speaking about the form of the challenges, is of the contrary opinion and says that in giving the denial, one must not immediately propose proof by arms, and that it is infringing his enemy's jurisdiction and doing the duty of the demander instead of being only the defender, that it is not appropriate that he both calls for a duel and has the choice of weapons as well. And on that, he concludes on this that seeing the fault done, the defender proposes the weapons [and] he must lose the choice of them. This (it seems to me) is believed strongly[32] and [I] cannot consent to this opinion. It is certain that a denial on an injury must be supported, and it seems to me that the denial the defender gives and the choice of arms which are available to him[33] are one, have the one and the same sequence[34] and cannot be separated. Since they cannot be separated, there is no reason that the defender loses the choice in saying "I will affirm it against you with arms."

Challenges must be made speaking in person[35] to the one who is the defender and must be made by a crier or herald.[36] For since by the King's permission, he is allowed to call his adverse party to a duel, he can order his crier to go call him out or, in the nearest village to his dwelling, in his own parish, in order that he cannot claim a single

---

[31] *amandé* should be *amendé*?
[32] *bien creu*
[33] *qui est en sa disposition*
[34] *vne mesme suitte*
[35] *à la propre personne*
[36] *vn trompette ou heraut d'armes*

excuse of ignorance.

Chapter XI – About seconds

The combatants must be very wise and prudent in their quarrels as to choose some brave Gentlemen, who are experts in arms, well known in acts of chivalry, and who can argue their honour, and councillors, together, to preserve the right which belongs to them. Also, this word (seconds[37]) is held for fathers, into whose hands they have been put. Also, the Seconds, between them, must be as loyal to their parties as they must be kept from being in any manner favourable to them, otherwise they may be greatly scorned for such a culpable act, in which they must be diligent to preserve the honour of their parties. In fact, in quarrels, one must resort to the judgement of those who have experience of them. Thus, after the Prince will have granted the duel to the two Chevaliers, they must each select a second, who will together visit the field[38] of the two combatants. The day and term of their duel having come, the second of one party and of the other will visit again the two combatants in order to see that they are not furnished with any spells,[39] helped by charms and other such manners of enchantment and sorcery, and will have made them make a solemn oath, one to the other – even the seconds could do similarly – and swear before God and the King and all attending that they have not nor are not accompanied by any enchantment nor want any such aid to their parties. Also, the parties must make an oath that they fight with good and just cause, formally declaring[40] before God and the King and all attending that they will only fight in just quarrels.

Chapter XII – About selection of the field

Several people differ on this opinion, namely, which of the two combatants must select the field. I was of the opinion that the defender who is called out to combat, having the right and choice of arms, must also select the field for their duel and make such selection that seems good to him. Since the attacker[41] is required to receive the weapons that the defender will present to him, is he also obliged to be found in the

---

[37] *parrains* – a term more usually applied to godfathers at a child's baptism
[38] *l'assiette du camp*
[39] *caracteres*
[40] *protestans*
[41] *assaillant*

field which will be assigned? And should the defender make known to the attacker the time, the day and the place of the duel? He could give any term that pleases him, be it two or three or six months, and, from all his terms, the attacker must choose the one that he wants and make certain of it for his part and assure him of not failing to be found there on the said day. By this, they will make ample demonstration that they do not want to flee the duel. And if anyone of the two wants to debate the disposition of the field which has been chosen, and has not striven to look for another, it would seem that he could be found a fugitive, which would be against his honour and would be greatly reprehensible to him.

It must be considered that kings be so sincere[42] and observe such equity in these causes of combat, where one debates honour, that they would not want to embrace the cause of one to defame the other. A noble King always follows reason and virtue and does not want us to hold one in different esteem, the duel thus having been granted by the King to the two Chevaliers. The seconds will be obliged to reconnoitre the field for equal distribution and sharing by the parties of sun and wind in order that neither one nor the other can complain of any advantage or disadvantage. For it is necessary in such combats to observe such great equality that the attendees themselves can judge the right and the reason which will have been made by the seconds. Otherwise, they may be to blame for having so indiscreetly put these two Chevaliers into the field. In short, it is necessary in this ceremony that there is no particular affection and that no one is so bold as to enter the field without the King's express command and of only leaving it [when] one has defeated his companion.

Chapter XIII – About the construction of the field and what it must be

The field for the two combatants must be far from houses and in a separate and special place in order to avoid advertising it to those who would want to give it [the duel] and also that no one should be leaning on the ropes. It must be constructed in a flat and clean place and without any hindrance which could prevent[43] the two combatants from fighting and stepping at their ease. And let it be traced and built twenty feet square or twenty-four at the most, and to the height of

---

[42]*véritables*
[43]literally, "harm" – *sans aucun empeschement qui puisse nuyre*

four or five feet and no more. It seems to me that this construction of the field is sufficient. To that end, the attendees can better see the duel which will be done there and speak of it with greater truth. The field is easily built of rope, even when the King permits the duel in his court and when it is in an army where the duel has been ordered. It is made between four stakes. I see it put up in these two ways: when the field is constructed of palisades, it is ordinarily for those who fight on horseback and it should be of greater space than that which I have just specified. It is only for these who fight on foot that I have written.

Chapter XIIII – Who must have the field built?

There are some who are of the opinion that it is up to the defenders, who have the choice of arms, to have the field built. Others say that it is the attacker who has the selection of the field. In this diversity, I opine that if the King orders the duel close to his person that he should have it built. For, since the two combatants have relied on the King's judgement, and that he wants to retain the definition of the duel, by this he demonstrates that it is he who must have it constructed. But if it happens that the King dismisses them to fight outside his court, I would say that the two combatants build half each,[44] or that the defender who has the choice of weapons furnishes them at his expense and that the attacker has the field built at his. I speak both for combat on foot as for that which is done on horseback. And regarding the weapons, they should be available to the one who must present them the day of the duel. And because in combat which is done on horseback, the costs are much greater, it should be that the seconds on one side or the other and friends who would be called on for this day find some workers in order to build it, to be paid after the fight and according to that which has been agreed by them and also to the satisfaction of the two combatants.

Chapter XV – Whoever touches the ropes of the list or the palisade must stand defeated

The ropes or palisade which have been put around the field have been put there with no other intention than whoever goes beyond them

---

[44] *les deux combattans le fissent construire par moitié*

or deforms them will be judged as conquered. And when ropes and a palisade are not available, it is necessary to lay a trench sufficiently deep and wide. Thus, instead of a trench, we put a rope or palisade there which must serve as if it were a trench and whoever touches the trench is certain to fall into it and by this means it would be easy to conquer [them]. Touching the rope or palisade as well, he must stand defeated. I say defeated if it were known to the eye and judgement of the attendees that the field rope would have been so strongly bent and deformed by one of the combatants that without arrest by the ropes he would have fallen into the trench, as much as it would be if the horse of one of the combatants had been forced and cornered against the palisade and had broken it. I hold both defeated. I ask what compels him who fights to touch the rope, even he who fights on foot? I judge that it is that which he retreats from and must consider that he would retreat further if he did not find the rope which arrests him and, in place of the rope, if there were a trench, he would fall into it. And if one holds his enemy against the rope, who is restrained by the strength of the rope [and] cornered with good thrusts, must he not be judged defeated? I believe that if one were to judge otherwise one would do wrong to the one who has thus fought against him. The lord Hierosme Muzio is of the contrary opinion and says that to have touched the palisade or rope, or to have taken one of the members out of the arena, for that, one is not defeated and that the battle should be pursued to the death, flight or recanting[45] of one combatant. However, he adds there that if there is no other capitulation made between them, since for who would capitulate otherwise, it may be necessary to accomplish all the points and articles of the agreement, under the penalty which may be contained there. However, I do not approve of this capitulation and say without capitulating that the combatants must not touch [or], I say, distort the rope or palisade of the field and, if my opinion is not found good by all, I declare to you that I say it only for those who would want to follow it. King Francis I allowed a duel in this Kingdom to two Spanish foreigners, one named Julien Rommaire and the other The Moor. They fought on horseback. Julien's horse having been killed, the Moor had no other trick than to gallop his horse all around his enemy. His horse being winded,[46] he was compelled to descend and, putting foot to the ground, Julien Rommaire promptly threw himself at him, putting him on the ground, sword at his throat. The King was present [and] said that these two

---

[45] *desditte*
[46] *son cheual estant hors d'aleine*

beggars[47] were not worth the effort of attending their duel. It was the same Julien Rommaire who was there when the army of King Henry II passed through the country of Liege, returning from Marienburg[48] which provided the location. I would judge that the duel on foot is much more worthy of the Chevalier than the duel on horseback. Also, it is the most common and from which the Chevalier must earn more glory and acknowledgement.

Chapter XVI – That the defender must enter the arena first with the weapons with which he wants to fight

The defender who has presented the weapons to the attacker must enter the arena first. Then he is named and must hold himself calm and still, awaiting his enemy. And as the attacker enters the arena, the defender must present himself and advance in order to prove his willingness to fight his enemy, not to wait for the attacker to go to seek him in the place and at the place where he first sees him. Doing this, he makes an error. But I say that as the defender will see the attacker come towards him, so must he be ready to step towards his enemy, and both are found in the middle of the field and begin to untangle their quarrel. But some say that if the defender sees his enemy is not ready to come to fight him, that he must hold himself still until he comes to attack him. I am not of this opinion because since both are in the list, they must not haggle to fight and [they] are engaged on their honour, so that they cannot flee without one or the other obtaining the victory. All ceremonies which must be observed in their quarrel and in their duel have been debated and settled before entering. And when they are there, they only have to play with the swords or the weapons which have been presented to them. I allow that it is up to the attacker to seek his enemy in order to draw out the reason for the wrong which was done to him. Also, it is on the defender to present this and to offer to satisfy him. He is not presenting himself when holding himself calm. It is necessary that he advance and come straight to his enemy. The lord Hierosme Muzio is of the opinion that the defender can hold himself calm and all that which he did before he saw his enemy on the way to attack him would be superfluous. If that took place, it would be like two statues that one has put in the

---

[47] *coquins*
[48] In the text, *Marzienbourg*

arena which did not move. For if the defender did not move from his place, until the attacker comes to attack him, and if the attacker does not go to seek him, are they not two persons that one has put in the list who resemble statues, since these two combatants have been put here to debate their quarrel and, regardless, they do not move from their place? However, they should both advance into the combat to which they have been ordered and that both of them have joined.

Chapter XVII – That the parents must not attend duels in the arena which are ordered

Relatives[49] must avoid the spectacle, which is too pernicious and certainly must be odious to all those who are present to see it, even if they are not relatives. Friends likewise must flee it. All the wise will fear to be found watching such a tragedy play out. Also, the play is not very pleasant and less agreeable to those who have judgement. They would willingly ask what satisfaction one could receive from watching his brother or close relative or intimate friend fight where before their eyes they know that he could give his life. I advise all relatives to not be found there where one plays with swords.[50] For it would be difficult, seeing his brother or close relative in peril of his life, that he did not give him some warning. It is more wisely done to withdraw from it. We will speak of this in the following chapter.

Chapter XVIII – That one must not speak after the Chevaliers have entered the arena

When the two Chevaliers have entered the arena, the King must have an ordinance announced that, on pain of his life, no one attending, whosoever he may be, be they brothers, relatives or friends, make any sign, be it with the feet or hands or words or by coughing that is made in favour of those who fight. Finally, such silence is required that all attending can understand whatever the two Chevaliers can do or say. And after this announcement, if anyone is found so reckless as to go beyond the ordinance that his Majesty has made, he deserves punishment by death. It is a fact which is of such importance that where

---

[49] *parens*
[50] Original text has knives but I cannot force this to make sense. Swords seems to better fit the author's intent.

one struggles for life and honour, there must be no acts of deceit. It is certain that he who does so would deserve a great punishment, even a shameful death, and claiming to be carried away by fraternal friendship cannot be admitted. Such excuses are not admissible. It would be much better to be absent from it.

Chapter XIX – About that which it is necessary the victors observe on the day of the duel

The Ordinance of Combat is from sunrise until sunset, and whoever does not confirm his word nor test it during this time can no longer be accepted to fight this quarrel. This is the opinion common to all Doctors[51] who have written about duelling. I wish nonetheless to dispute this reason. Not that I want to change their opinion, for I know that they have such good knowledge that I can only follow after them, and I will deem myself knowing enough when I follow and imitate their beautiful doctrine except for clearing up this little matter. I will propose a question, namely, if two Chevaliers have not triumphed at all during the whole day, neither by death nor by wounding, such that all attendees recognise the brave valour and singular boldness in these two Chevaliers, who have not spared their lives [and] night overcame them debating valiantly their quarrel, I ask the judges which of the two must stand defeated? Next, it is necessary that I give my opinion on it. It seems to me that the duel should be postponed, if the Chevaliers are of this mind, with the King's permission. These two Chevaliers, who have entered the arena, are not separated from each other by their own wills but night has put an end to their duel. I say there above that they have used their life. And if the attacker has dealt with a valiant man, he is not to say that the defender is not so valiant, since they have made such a test that they [both] must be praised and esteemed. And for this reason, neither one nor the other must stand vanquished for that day but must both be judged equal combatants because both have debated their quarrel and have satisfied it valiantly. And if the King judges the combatants equal, they must be content and no longer enter into combat for this quarrel because they are held in the same honour. It seems to me that I am not inappropriate[52] and

---

[51] We assume here that the author means Doctors of Law, although this is not stated.
[52] *hors de propos*

believe, if my opinion is admitted,[53] one will find there something approaching reason. I will pass over and will come to the time the duel has ended. He who will be victorious must remain the last in the arena and his enemy must be the first to leave. The conqueror must call to the King and all his attendees, if his enemy returns to hear his enemy's words, in order that he and all his company can bear good witness and, at that, the conqueror will say to his adversary that he puts down the weapons and that he comes to him. Being come, the second of the conqueror will disarm him, then the conqueror will lead him out of the arena and will request his Majesty do justice by him for the wrong which his enemy did him. Thus, the form which I would that was kept on the day of the duel of the two Chevaliers. And in order to make it more clear still, whoever is victorious speaks freely in these terms: "Give me my honour, life and arms." It is therefore reasonable that he [the victor] be disarmed before he leaves the arena and when, for his life, he leaves it at the will and discretion of the King. For even though the adversary has been given to the defender or the defender to the attacker, he cannot for this kill him. This would be an act of a tyrant for which he must be greatly blamed, even punished and chastised.

Chapter XX – Of the choice and selection of weapons

All those who have written on duels have conferred the selection of weapons to the defender and say that there is reason, especially as the attacker is obliged to prove [the charge], and that it suffices that the defender defend himself well. I know that one-on-one combat in the arena will be more perfect and better esteemed by Chevaliers when they desire to finish their quarrel with thrusts,[54] without any selection of weapons other than sword and dagger or sword alone, which are the Chevaliers' weapons. That the defender is obliged to provide [them] of equal length and that the aggressor will choose whichever of the two as pleases him, I well know that I will be told otherwise. That the sword alone still is in use, that this combat is the most prized by valorous Chevaliers, it is nonetheless honourable combat and it re-

---

[53] *bien goustee*
[54] *en pourpoint* Confusing. A *pourpoint* is a doublet but this makes little sense in context where a term like *chemise* (shirt, shirtsleeves) would make better sense. However, the *Dictionnaire du Moyen Français* records the noun as a derivative of the verb *pourpoindre* which carried the figurative sense of to poke, to stick, to needle.

sembles the beautiful and furious savages who with rage and great ferocity entangle themselves in it without conceiving of death. The King must forbid this manner of fighting with the sword alone, even if this is the only [form] that the two combatants agree to. For one knows well enough that there are those who are more fit and more apt to fence than others,[55] and if one is more experienced in fencing than the other, there would be great expectation that he could carry the victory even though his right was neither good nor legitimate. This would be something similar to a gladiator calling out one who had never handled a sword. Anyone could judge that the gladiator must obtain victory by reason of his long experience and because those of my opinion [and] others [who] do not agree, let us speak of the arms which should be chosen and which are more appropriate and which are to be rejected.

Chapter XXI – About the proper weapons for the duel

We have said above that the sword alone is the most used for duels and is the most common for Chevaliers, although it is practised differently today, and all those who argue for it want to maintain this opinion. I want to agree with this. I will say, then, that the Chevalier who is the defender must present to his opposing party weapons which may be judged worthy of combat. For if he gives unusual weapons[56] of the sort which cannot be handled, these weapons are not appropriate, nor should they be accepted. He should produce them that may be of such justification[57] that they can be taken and held as reasonable by the Prince, firstly, [and then] by the seconds and the attendees, and that they may be made according to the disposition of the two Chevaliers and their bodies, for if one is left-handed, it is not reasonable that he constrains his opponent who is right-handed by nature, to fight with the left hand. It is required that both fight according to their nature. And to say that one can be as well aided with one hand as with the other, there are right-handers who do not know how to do anything useful or skilful with the left hand. Also, providing *brassards*[58] which prevent bending the arm and *cuisserots*[59] which keep the knees from

---

[55] *qui font plus à dextre & plus a droits à tirer des armes les vns que les autres*
[56] *armes non vsitees*
[57] *qui soient de telles deffences*
[58] Armour for the (upper) arm.
[59] Armour for the thighs.

stepping, these would be weapons of deception. But, if the right-handed by nature were crippled in the right arm and for this reason was made to use the left, I would be of the opinion that he must constrain his enemy to fight with the left hand like him, and could provide him a brassard which could prevent him using the right arm. Also if he is limping[60] and, with the same member in which he is crippled, he can spoil his enemy. Also, one who had lost an eye provides him with a helmet[61] which hides his own, without which there would be certain fraud. But, if the two Chevaliers are healthy and of the same disposition, same age and same strength, with what weapons do you want them to fight? Moreover, if the defender presents weapons which constrain his enemy, it is to take and seek an advantage over him. Since you see today powerful forgers of weapons which are neither reasonable nor in any way usable, it would therefore be necessary for me to answer that the defender should present defensive arms and those that one knows to be the most honourable and that Chevaliers will be most often used to and accustomed to fight with in times of war, in which the seconds must be curious to observe that, in the selection of arms on the day of the duel, none may be presented that are not worthy of being received, and worthy of an honourable Chevalier. And, if the defender on the day of the duel presented those which are novel and unusual, and that the assigned day of the duel passed in this dispute, I would be of Sir Muzio's opinion in this, who says, "the passing of the day without combat through the fault of the defender must run to his disadvantage and that the attacker has satisfied his duty." In combat, one uses ties[62] and wrestling, things of which I cannot approve. There is no doubt that there are those who have this disposition and more bodily skill than others, and who are confident in their dexterity to promptly throw the body of their enemy with strength or skill to the ground, by which they obtain the victory. This is not done through a brave and valiant means of fighting but rather by trickery and I know only that it is the disposition of the body which surprises his enemy. I cannot esteem this combat. Since there is a selection of weapons, one must only fight with [these] weapons and I believe that those who do otherwise must not be taken for valorous Chevaliers, for throwing his enemy to the ground is not to fight him. It is to wrestle like a pile of beggars who fight with fists and scratch the face. The nobility[63] of the valiant Chevalier is known by his vow and his brave

---

[60] *boitteux* - limping, lame, crippled in the legs

[61] *bourguignotte*

[62] *liaisons* – short lengths of cord binding the opponents together

[63] *la magnanimité*

courage, wanting to undertake nothing which would come back on his honour and all that which he does, be it in war or single combat, he desires will end in his glory and honour, in order to be recognised for a brave and valorous Chevalier. The Chevalier is not considered valiant if he fears anything. It is necessary that if he wants to acquire a reputation that he conducts his valour to so happy an end that he may stand in honour and be recognised for such. I want to say therefore that he must end his fight with weapons. And being of this opinion, that when the two Chevaliers enter the arena, it should be declared that neither of the two Chevaliers will take the other by the body. I would gladly ask, if this had happened, would it serve for the choice of weapons and of the sword? I know that ties have been used in all times, and all have approved them. But I cannot approve of them, even in the particular arena where honour is being debated and where one must prove his valour. In this way, most often, has he who is wrong obtained victory. Here is a very great cruelty to which the King and the seconds must have a regard. I have nonetheless seen in the army of my lord de Guise,[64] that he had in Italy for King Henry II, being before the city of Cynitelle in the country of Brousse in the Kingdom of Naples,[65] the duel of a captain of infantry[66] and his ensign, which occurred in the city of Cynitelle for some difference that they had together and asked my lord de Guise for a field, which was granted to them between four pikes. The ensign seized his captain by the body thinking to be the strongest, however, the captain, named Semerille, held him so firmly that he constrained him to surrender and saved his life. He had another in his camp close to Rome. Captain Aprouillan, also from a company of infantry, against another Italian captain, who, being inside the camp, threw himself at Aprouillan, who was weak and was defeated. I speak of it in the manner of advice given to me by those who have the experience and know more in order to judge it by that which they know to be good.

---

[64]Francis I of Lorraine, 2nd Duke of Guise, 1st Prince of Joinville, and 1st Duke of Aumale, b.1519, d.1563), was a French general and statesman, a prominent leader during the Italian War of 1551–1559 and French Wars of Religion, assassinated during the siege of Orleans in 1563.

[65]Likely Cimitile; the country may be referring to Abruzzo (there are two in the Kingdom of Naples: Ultra and Citra [upper and lower]). Geographically, though, Cimitile is just outside Naples proper and would actually be in Lavora, though Lavora borders Abruzzo Citra.

[66]vn capitaine de gens de pied

Chapter XXII – If the Prince can Interrupt the Duel

It is something that does not happen often, if at all, that the King separates the two Chevaliers from combat, when once they have entered the arena. It is the common opinion that they must be left to complete the duel and several think that if the King separates them that it would be against the obligation of his own promise. I respond to this that the King, having allowed them the duel, and the combatants, being entered into the arena and debating their quarrel, I believe that at this point, he has satisfied his promise and that the two Chevaliers are equal in valour and that both are brave and valiant. Not wanting to lose two very brave Gentlemen, being accompanied with the honest duty that God commands Princes to observe in the place of the Chevaliers, that they are acts sufficiently pitiful that to save such valiant men, when he separates them in this manner, I judge that these acts are virtuous and very laudable, to command them through his guard, with the advice of the two seconds, to lay down their arms and hold them away from them. Likewise, with no longer fighting over the subject about which they entered into the arena, and that they both have occasion to be content with proving that they have done it but, if one of the two is wounded and the other is not, he would do wrong to him who is not [wounded] to separate them before he had known his opponent wounded. Then, they could be separated.

Chapter XXIII – If the two Chevaliers, being in the arena, can change the quarrel

I believe that if one of the two Chevaliers, being in the arena, said to the other that he is villainous[67] and that the latter responds to him with his mouth that he does not intend to fight on the first quarrel but, of course, on former who called him villainous, giving him the lie, I judge that in this they have not satisfied the first quarrel. For it is not permitted to change the quarrel if the first, for which the arena were granted, is not completed. But in order to change the quarrel for this, I would not want to argue that he was defeated. He should wait for the issuing of the second [quarrel] and if he is victorious in both, then I would say that he who wanted to change was dishonoured. If the one

---

[67] *meschant*

[combatant] was victorious in the one [quarrel], and the other [combatant] in the other [quarrel], I would hold them both for villains and for honourable Chevaliers. Sir Muzio is of the contrary opinion. His reason is, because they undertook to fight for one unjust quarrel, because the victory in the one does not relieve the loss of the other, they can be reputed as defamed in all other quarrels. It is his opinion. I want to neither condemn nor approve it but if the Chevaliers change the quarrel for the same topic, I believe that there would be great evidence that the quarrel is just, as one that I have heard told, which will not be inappropriate to put in these arena, in order that afterwards those who have read it can make some judgement. Two honest Gentlemen,[68] of brave valour and very great friends, having only one conversation, conferring together privately, one says to the other that he has lain with a woman whom he names [and] the other repeats it. [The former] complains of the dishonour that this Gentleman [the latter] has done to him by retelling their private conversation. She enters into the conversation with him [the former] with fine insults. He responds to her that he has never spoken of it and that he who had reported such words has lied in his throat.[69] He who lied [the latter] called out the other to duel and demanded that he will maintain in this arena [the former] having told him. He who had given the lie [the former] said these words to him before all attendees, "I want to enter into a proof of arms with you to uphold to you that you have done a villainous turn by having retold that which I had told you as my faithful friend and, having done such a wicked act to me, I will thrash you liberally and try to have you killed." They fight. The Gentleman [the former] who had said the words to his friend and dishonoured this woman was defeated. I say that this quarrel was well debated and on the same topic of the quarrel which was between these two Gentlemen. This is not a change to the quarrel since it was on the dishonour of this woman that they fought. For when he said, "having retold that which I said to you," by this he avowed that he said it and when he said, "you have done me a villainous turn by having retold that which I said to you as a faithful and private friend," by this he changed the quarrel but it is on the same topic, in which I find that it is well-founded. Yet, is it not a proper change of the quarrel? He declares to him but he adds that he has done him a bad deed in having retold it. When a friend tells you something of importance in secret, it must not be retold so that it not carry disgrace to anyone in retelling it. Although the truth is

---

[68] It is almost impossible to untangle who the various pronouns in the original text refer to in this sentence.
[69] Idiom?

that he said it, the truth is that you have also done him a villainous act by retelling it. I praise greatly that when one would change the quarrel that one not declare it by words but, without speaking, that one uses some artifice where there was a showing of reason and honour all together. As the two Chevaliers, being in the arena to debate their quarrel, he who thought he did not have the right, turns his back and makes seeming to flee, the other shouts at him, following him, turns coward. The Chevalier who fled returns and kills his adversary. In fleeing he made a demonstration that he did not want to fight on the first, yet he did not explain it, and in returning, he had recognised that he fought on the injury that the other did him in calling him a coward. He was correct in both. In the first, he did not want to fight it because he knew that he was wrong. It's the reason why he made a show of fleeing. In the second, he turned because he had been insulted and to prove that he did not lack courage. Monsieur Paris, in his book on combat, recites about one who, fighting in the arena, cries out, "I surrender" and in the same tempo killed his enemy. Why, he concludes, should one have regard for the deed and not the words? I abide strongly in his opinion.

Chapter XXIIII – The rights which belong to Chevaliers after victory in a duel

All the weapons of the defeated, by the institution of duelling, belong to the victor. If he surrenders to his enemy, he should similarly give up his weapons. If he is killed, his enemy can strip them and carry them away as being the true marks of his victory and should not be prevented from this, otherwise it would be wrong. Lord Muzio makes a strange speech about the person of the defeated and says that he must stand a prisoner of the victor and that this must not be refused to him by anyone. He can demand a ransom and the defeated can even be used by him and, if he cannot satisfy him with the ransom, serving the space of five years in work suitable for a Chevalier, [after which] he is free, without one being able to ask for payment for his upkeep. He proposes many other conditions which are too long to deal with. I will endeavour to respond to them. First, I will say that the defeated must not stand the prisoner of the victor because the Chevalier who fights for honour aspires to nothing other than obtaining victory, [and] obtaining it, is this not sufficient advantage to him without holding the defeated and making him pay a ransom and costs which have been incurred by reason of his duel? This would be to make duel between two honourable, valiant Chevaliers similar to one as if it were two peasants who were fighting a duel one-on-one in the arena, which is permitted only to Chevaliers who feel insulted and who have honour and reputation to preserve. They must, therefore, finish with an honourable end, which is victory with the reparation of honour. Thus, that which I have to say on this article. Let's now come to that where he says that the victor can constrain the defeated to serve him. It is a strange opinion, being served by a Chevalier. I would ask in what vocation he would be able to serve. He will not make him his stable boy[70] nor his cook. [These] offices are too lowly. And if he served him in these qualities, he should therefore negotiate the business, but this is not permitted to him, especially since he is a prisoner. I will make a comparison to two Chevaliers who have a quarrel. The Baron Tevoir has a quarrel with the Baron de la Saulay. They enter into a proof of arms. The Baron de la Saulay, who is rich and lord of twenty thousand *livres* in rents, is defeated. The Baron Tevoir is the victor who has no other possessions or means than three thousand livres in rents. Should de la Saulay be a servant to Tevoir? Thus, a Gentleman of a good house would be badly treated. This would be an act to disgrace not only him but all his offspring[71] and a cruel and barbarous duty. In the example of Tamerlane, who defeated Basajet in a pitched battle,[72] he put him in a great iron cage and had him dragged after him in order to serve him as a footstool[73] when he mounted [his] horse. Of course, I shall never be of this opinion and there is neither law nor reason which can prove to me that this act must be done between Chevaliers of honour. Perhaps this ordinance, that he takes prisoners, may be good in Italy and in some places, and not in all. But in France, this is cannot be practised nor suffered, and the King, the Princes, his council, and his nobility will not permit it as a thing which is unjust and done without reason or judgement. If this took place, it would only be necessary to stand a gallows at the end of the arena in order to hang the defeated. One should not seek greater shame and greater punishment from the defeated than that which the ordinances of the duel command for him.

---

[70] *palfrenier*
[71] *toute sa posterite*
[72] *en bataille rangee*
[73] *marchepied*

Chapter XXV – Of he who will have accepted the duel but is not found on the assigned day

When the King has permitted a duel to two combatants, it is necessary for their honour that they do not fail to be found in the assigned place, and when one of the two combatants will make that fault of not being found on the day which has been given to him, dishonour and cowardice must be attributed to him. Thus, why the Chevalier of honour and valour must rather choose death than endure such a reproach if it were not that he has fallen in an extreme necessity of illness, which constrains him to keep to his bed as unable, that he is completely exempt from combat, and in this case it cannot be noted as disgrace. But, being healthy and well disposed, he cannot be excused. The Chevalier should nonetheless make a show with his excuses, which may be good and legitimate, so that we know there is no fault[74] in his motive. There are many kinds of hindrances which happen inopportunely and which must be received by Chevaliers so one must put them to the judgement of the King, his council and the seconds in order to advise upon according to their worth of being received. It had been the custom that if the Chevalier who was called to fight was not found on the assigned day, his enemy could have his coat of arms[75] dragged through the streets or around the arena as a triumph, and beseech the King to remove his arms and nobility from him. This opinion must not be received among brave Chevaliers. But I may well be of the opinion that if he who should not be found in the arena on the assigned day, his adversary could have him called by trumpet knowing that if he is not willing to uphold and defend with weapons that with which he is accused, and must wait until the sun sets, truly a failed day, and stand in the arena, and beseech the King to do him this honour, since his adversary has failed on the day and that he has not made known his news of having had justice on him, his Majesty must declare he who is in the arena victorious, his adversary unworthy of wearing the sword and, if he has any rank or honour, [it] must be removed from the one who does not deserve them. However, he can put him again in his honour by doing some service to the King or to the country, in consideration of his nobility and the house from which he has come.

---

[74] *la faute ne soit venuë*
[75] *ses armoires*

Chapter XXVI – About he who has been defeated once in the arena and, after being called to a duel by another, stood victorious, if he has by this recovered his honour

The lord Hierosme Muzio in his book on duelling proposes this question and opines, "that one must hold one maxim that the loss that one had on the first duel is not restored by the victory that one has in the second." And on this question, he argues the advice of Alfonso d'Avalos, Marquis of Vasto,[76] who said that he who left the arena defeated, shows that he had held his life higher than his honour,[77] and that if he enters once again into a proof of arms he repairs his honour because of it, especially as it is presumed that he tested this fortune in order to determine whether he could remain the victor, with care, nevertheless wanting to save his life at all costs. Such is the judgement to which Muzio concludes and maintains for chivalry. I will respond to this question and add here my opinion in order that those who will read theirs and mine can make some judgement and follow that which they understand to be the best and the most certain. Not that I want to contradict the opinions of lord Muzio, nor that of the Marquis of Vasto, since I also hold them chivalrous, and know that lord Muzio wrote better about the duel than any living man could do, with such good instructions. But in France, duels and the honour of the fight are conducted entirely otherwise, differently to other nations, especially as France is filled with nobility who make profession of arms, from which the King draws very great service, and for one who would have overcome a disaster or some other accident in combat, for this, the French Gentleman does not stand dishonoured, Fortune being such that he will be found instead where he will recover altogether the damage which he did and his honour. In France, the Gentleman continually has weapons in the hand, whereby such will have the debate with one [against] whom he will stand victorious, and after, he will take the debate to another and lose the victory. Why does he value one or the other combat? In the first, the weapons laughed with him. In the second, they were unfavourable to him. And in both duels, he has proved his valour. I propose an example: Let's suppose the

---

[76] Alfonso d'Avalos d'Aquino, 6th Marquis of Pescara, 2nd Marquis of Vasto, b.1502, d.1546. An Italian *condottiero* from Aragon, renowned for his services to Charles V, Holy Roman Emperor and King of Spain.
[77] *qui la plus fait conte de sa vie que de son honneur*

case that Scipio and Hannibal have a quarrel and go into the arena to finish it. Hannibal leaves the arena defeated. Scipio is the victor. Scipio, at some time afterwards, undertakes another duel against Marc-Anthony where Scipio loses the victory, and Marc-Anthony stands victorious. Should Scipio be dishonoured for this second duel and lose the honour which he acquired in the first? I assure myself that all the Chevaliers who understand the exercise of arms advise no, and that he can enter into proofs of arms against all those who would call him out, notwithstanding the loss which he had in the second duel. Now, let's return to our example in order to make a more evident proof of my words. I said that Marc-Anthony remained victorious over Scipio, who had won against Hannibal. It happens that Hannibal has a debate with Marc-Anthony [and], having entered the arena, Hannibal is victorious [and] Marc-Anthony stands defeated. In this duel, has Hannibal not recovered his honour and erased the loss of the first combat which he had against Scipio? I say yes. I will say therefore that all Chevaliers who have the heart, boldness and confidence to enter the arena in order to debate their quarrel have made sufficient proof of their valour and of their bold courage, and are not presumed to enter the arena in order to fight and only try to save their lives. But there is much more to the appearance than making a brave feat of arms. Rather they expose their lives freely to such peril that it could happen, since it is not likely that a Chevalier enters into a proof of arms with sword in hand in order to save his life. It is to fight or kill. And then I ask why the situation must not be as good for him who is defeated in the first duel as for he who conquers and then is defeated? It seems to me that it must be the same. I know of no brave Gentlemen who have fought and have been to a war in which he fought who have not been wounded. Should it be that they stand dishonoured and no longer carry weapons? This is not reasonable and the opinions must be better examined. I advise that he who was defeated in the first duel must not be for that refused a second and, if he stands victorious in it, must be held as a brave Chevalier. When a Chevalier is declared incapable of no longer entering the arena to fight, it should be done by the express commandment of the King, which will be difficult depending on his lineage and posterity. If it is only the accusation that one made about him, be it proven, still it is necessary that it be a capital [offence] and worthy of death. Nonetheless, his Majesty can, if it please him, give grace to the defeated, enjoining him to do him an indicated service in order that all the evil which could have been done be effaced by some brave feat of arms and that he proves his willingness to be of service to his King. This is the custom of this Kingdom and

I have seen the same judgements given from the King's mouth that I find good and very reasonable for the Chevalier. It is a maxim that when a Gentleman has made a stain on his honour, he should repair it by arms as soon as the occasion presents itself in order that he makes known that which happened to him is not for lack of courage or valour. There are truly persons who are happy to duel and who favour weapons, others are accompanied by unfortunate happenings,[78] and nonetheless are valiant and bold. For this consideration, they must not be rejected from the company of Chevaliers.

It is now necessary to more fully examine the judgement of the Marquis of Vasto in which there are two points to note. First, when he says that the Chevalier, even though he enters into proofs of arms, [as] one has in former times and that he may be victorious, not withstanding this which he must say, that he has repaired his honour, understanding that one can presume that he may be found there with the intention of testing Fortune in order to see if this day he can remain victorious. The other, when he says with intention however wanting to save his life, cannot attain the honour which he had, having once completely lost it. Thus the proper words of his sentence, where I say that there are two points which repel and are contrary to each other. It is a general maxim that two contraries cannot stand together in the same topic, such that of necessity there should be one which erases the other. These two points, that have come from the opinion of the Marquis, are opposites: that he may not thus enter the arena in order to test his fortune if this day he could be the victor. To test Fortune is to risk one's life or to try at the peril of one's life to obtain victory by risking one's life.[79] It is the duty of a valorous Chevalier to acquire much honour. So it must be concluded that a Chevalier, having a lot of valour and honour and who valiantly risks his life in a duel for his honour, must not be denied nor refused the privilege of a duel, even if he had been defeated previously. The other point [is] when he says "with intention however wanting to save his life." These are words and effects which are complete opposites, to fight at the risk of one's life in order to want to save it is something which cannot be done. One knows enough that the event of a duel is uncertain, and if it happens, it is through fate and the will of the victor. There was a duel in the arena in France between two Chevaliers. The victor cut the hamstring of the defeated.[80] Being on the ground, the victor asked him that he surrender. The defeated responded to him "I will not sur-

---

[78] *quelque sinistre aduenture*
[79] He seems a little redundant here.
[80] *le vainqueur couppa le jarret au vaincu* Sounds familiar!

render. Kill me." That the victor does not want to do [it], should it be said that he [the vanquished] had entered into a proof-of-arms to save his life? There would be no appearance of reason in it when he said to him, "Kill me." Thus is it with all the brave Chevaliers who enter the arena in order to debate honour. They also endeavour to maintain for themselves a brave reputation and such that the Chevalier of honour must acquire. If he was not killed there for that [reason], must we conclude that they have made more account of their lives than of their honour? It seems to me that it suffices for me to have debated on this opinion. I will allow others to speak of it who can understand it better. I will content myself with them, having opened the way in order to speak about it more fully.

Chapter XXVII – If one must be received to fight the quarrel of another in the arena

All those who have spoken of duels are of the opinion that when a Chevalier is offended by a lesser, he can then present another in order to fight the quarrel. Or, if he is sick or that if he is in his minority and not in strength to dispute his quarrel with weapons, they name for themselves a champion: some call them combatants, others respondents (which seems to me to be the most appropriate). Since like for a debate, one gives a surety[81] in order to respond and to be obliged, and to him alone can one attack. Also, he who responds in absence or in presence, be it for he who is in minority or feebleness of age, is obliged to respond in proofs of arms and to fight his quarrel. It is always necessary to regard this question and advise if the champion that the Chevalier will present is worthy to respond to the adverse party, if he is of his status and if he can be compared to the Chevalier who wants to fight him. And how much the Chevalier wants to debate his quarrel with a champion or respondent who is not of his status or his equal or in any way similar, he must not be permitted by the Prince, together with all the seconds, for that reason. The Chevalier of honour must make a proof of arms with another who is equal to him and not otherwise. For if a lesser than he in name and arms and status, or a yeoman overcomes him, he receives a double reproach: the one because one might have judged that he would not have fought in a just quarrel, the other that he had been defeated by a small man of low status. I will say nonetheless that however great a lord he may be,

---

[81] *une caution*

he will never have honour by presenting a champion for fighting his quarrel against anyone if he is not a Prince but only carrying the title of Gentleman. He cannot refuse another Gentleman, even though he is not the same in status and, if we find there such a great disparity, the King in this can use his absolute authority in order to grant them their quarrel and not to suffer that he fights through a champion. I am of the opinion that a Chevalier will never have the right to fight the quarrel of another if he does not know it very just, and believe if he does otherwise that God will never favour this duel. I have read some histories which are appropriate to be put in this place. It was that two young Genoese Gentlemen, coming from the Isle of Cyprus, being both in the same boat, one was named Ottobon and the other Grillo, took up a quarrel. Ottobon was no longer seen and it was believed that Grillo, who was the strongest, had thrown him into the sea. Ottobon's relatives complained about this to the Magistrate. On this complaint, Grillo was made a prisoner and, because there was no sufficient proof, the judge commanded that the parties elect their champion in order to come to a duel and put it to an end. Ottobon's relatives selected a Florentine called Caccica and, for Grillo, one named Pistello de Como. Both being entered into the proof of arms, the Florentine stood victorious and the judge, according to the law that the Lombards had introduced, sliced off Grillo's head as guilty. He who will present himself to fight the quarrel of another must be the son when the father is old and frail or not having made the practice of arms, being occupied in private and small affairs of his house. The father, too, can fight for his son when he is weak of body and of age. The husband for the honour of his wife and the wife, lacking a husband and son, can present a combatant and respondent to debate her honour, provided that it be a Chevalier of honour against another Chevalier. The brother for his brother and for his sister, relatives close in consanguinity, like those who want to participate in the quarrel of their parents. The Learned[82] who wrote about duels say that the circumstance can happen that the lord is constrained to take up arms in person against his servant and show reasons, and similarly for the servant to the lord. I would not consent to this opinion that the servant can debate an individual quarrel with his master, and do not see that one could build a quarrel for any occasion whatever. A servant is a serf who has no status and is completely unworthy of fighting his master in the arena. Laws condemn the vassal as a felon for undertaking against his lord and in this regard the lord could confiscate the fief of his vassal. Such quar-

---

[82] *Les Docteurs*

rels when they occur must be ended by the courts[83] and not by arms. The King must condemn [those] following this track and maintain the lord in his rights when the vassal will break away from his duty.

Chapter XXVIII – The ways of giving the lie

This matter of giving the lie must be well understood, for the difficulties which are found there and for ignorance many persons, having occupied themselves in writing on this question, have not found it too difficult and several have spoken about it very diversely. I will pray the reader follow the opinion of those which he finds the most reasonable. In order to follow my intent, there are several types of giving the lie; some are certain, some are conditional, others are general. The certain are those which are said with assurance of having seen, heard or heard said because there is no more certain witness nor a more true testimony than that which the eye has seen and the ear has heard, and to which we must pay closest attention,[84] even when it issues from the mouth of a man of means and from a Gentleman of honour. And when the contrary is objected, they can give the lie, and this giving the lie is called certain. As, for example, if I address myself to lord Fabrice, bearing to him such words [as] "Lord Fabrice, you said in the company of several lords that I was a traitor to the King and that he must not trust in me. I maintain to you that you have lied." This lie is certain and particular, particularly as I speak to he who spoke badly about me, and certain because it was witnessed by notable and worthy Gentlemen of truth that he maintained such words about me. For if one has not sufficient proof of that which one objects of to his enemy, the lord Fabrice in this case could be well founded to give the lie and to prove that he had never said it. But also, his adversary, having proved it by witnesses notable and worthy of faith, can always give the lie, that which one will call a certain and legitimate lie, and Fabrice can only enter into the arena to debate the contrary of this lie. All the valiant Chevaliers must use and hold it as true and legitimate, without going to seek such untoward crossings in their quarrels and of poorly giving the lie, where very often there is no basis, which very often turns to dishonour them. We will speak now of giving the general lie. I say that when one gives the lie in general terms, that the lie must not be received, as for example, "whoever says that I fled from the battle has lied." This lie obliges no one to respond to it be-

---

[83] *la iustice*

[84] *à quoy nous nous deuons le plus arrester*

cause it is not legitimate. There is another general lie. For example, "Lord Paul, you have spoken badly of my honour. For this reason you have lied." and because he has not specified the words which he felt offended his honour, this lie is called general, and [I] counsel all good Chevaliers when they would repulse the injury that one would have done to them that they do not use this lie. Since, if it happens that Lord Paul responds yes and produces for him another act which he has done, about which he will submit to have him prove, thus, Paul could say that it is he who has lied. Thus why it is necessary to guard against giving the general lie, and always name the injury for which one intends to give the lie. We speak now of giving the conditional lie. These are those which are given with conditions. For example, "if you said that I was a thief, you have lied about it." To this lie, one could respond, "If there is a man who would accuse me of saying it, he has lied," or find some other subject in order to deflect this accusation. In these responses, it is necessary to be wise and well advised. For the Chevalier who is careful with his honour must be so well conducted that he may neither be too prompt and too rash to advance words nor give the lie which could make him fall into greater error and must always maintain himself with honour and right on his side.

Chapter XXIX – Of the form and manner of giving the lie and how the Chevalier must conduct himself

I have always been of the opinion that most quarrels which are made today happen due to[85] mockers and backbiters, and that it greatly displeases a man of honour to see himself mocked and derided in good company, meriting the answer from which giving the lie is born. This insolence must be chastised and reprimanded by the King and by those who can have known it, and if it touches the honour of he who is offended, he would be constrained to proceed to weapons. This word "lie" should be repressed. Also, I have never been of this opinion that such quarrels were neither better nor worse than those tolerated by the King, and [on] any accusation that could be made, the defender would never have the honour of giving the lie. It suffices when he is accused of some crime that he responds that it is a false and imaginary thing, and that he will uphold it with arms. This is to respond with honour in his quarrel. I very humbly beg his Majesty to have

---

[85] *procedant à l'occasion*

an eye so that the lie may not be so lightly given and make express ordinances about it and a general edict that between Gentlemen, one has to complain today [that] the lies are very familiar between Gentlemen, as with the porter who complains about everything without accepting the consequences of their quarrel's merit, and even less of their person and status, and that one calls out another for justice in order to have reparation for a lie that one was mocked with, even between Gentlemen who make profession of honour. King Francis I said one day in the presence of many great lords of his Kingdom that he who endured a lie was not a man of worth. He said it because Emperor Charles V had said many bad things about him, to whom he had complained through his heralds. But that this speech left the mouth of so great a King was the occasion that even the valets killed each other in order to give the lie and this made a great slaughter. This is why the King must abolish all giving the lie and make a belief which carries a penalty, as I said above, in order that there is not one in his Kingdom, having permission to carry a sword or not, who would dare deceive another, and he who would deceive one (here this is at an insult which being proved could make the one who would have received it disgraced or worthy of death) must make himself repair his lie, and slights which are said in anger are repulsed with the softer denial and without giving the lie. For the Gentleman making a profession of honour will have more honour repulsing the insult which has been done him with the sword than with a lie. There are some who are so unconsidered in their language that being well accompanied finding their enemy alone, they give the lie to him, this lie must be as nothing, being given with advantage.

Chapter XXX – About the ambush[86]

When someone takes another at his advantage, this has always been held as an ambush and the act of a coward, like if someone wanting to take advantage of the injury one did to him and being well accompanied and the stronger comes to attack his enemy. This must be named as an outrageous ambush made by one who has insufficient honour and virtue to be seen among valiant men of courage [and] to demand face-to-face[87] satisfaction for the injury which he had re-

---

[86] *supercherie* - Cotgrave (1611) give this as "foule play; an iniurie, wrong, affront, bravado, assault on a suddaine, or upon great advantage." The *Tresor de la Langue Francaise* gives it as *tromperie* (attested in 1611) and an *attaque par surprise* (attested in 1566).

[87] *cap à cap*

ceived but, on the contrary, gives very evident proofs of his cowardice and of the little courage which is in him. He who has been attacked in this fashion must not stand dishonoured even though defeated. For who is he who can guard himself from being taken in treachery or disadvantage? I believe that it is impossible that one can guard himself from the knavery[88] that a wicked man wants to do. And if it was that someone wronged his enemy or another to his advantage, one could respond to him "if fortune had favoured me such that we met one-on-one, you would be kept from offending me, that if you had the confidence to present yourself alone, and that I do something which abrogates[89] the duty of an honest man, you will have the occasion to hold me for a coward and without courage. I entreat all the company who is with you to hear my words in order that they can give witness to that which happens today between you and me." And every time he who is offended is obliged to take it up, even at the price of his blood and of his life, and to try to have satisfaction by arms, and to beg the King instantly to permit him a duel, by holding his opposing party for a Chevalier cowardly at heart and unworthy to carry a sword, who does not dare to call him out one-on-one but who has outraged him at his advantage. It is indeed required when one such dispute falls before the King that the motive of this quarrel be well examined and this cause prudently be determined by his council. It would be a regrettable thing that the honour of a kind Chevalier[90] was engaged by the rashness of one who is without judgement and without discretion but who, swollen with arrogance, is not content with insulting a man of worth, an honest Gentleman, but goes against the proper process with advantage. For this consideration, the King, before according the duel, must be informed if the quarrel merits a duel or not and whether there is no other means of seeing reason than with the sword, since I believe that if the act is done basely and by a bad Chevalier that his Majesty will judge this case should merit an exemplary punishment, or of making an honourable reparation, and think that one would do a great wrong to a Chevalier of honour and honourable reputation to force him to duel one who will have so grievously and cowardly wronged him. And because it is a dispute which happens sometimes between Gentlemen, the King must take recognition of this cause, and take it up so adroitly and with such affection, that it constrains the one who would have cowardly and to his advantage offended a Chevalier of honour to make an honourable

---

[88] *meschanceté*
[89] *desroge*
[90] *gentil Cheualier*

reparation for it, and constraining him to say to him that he offended him at his advantage and admit that it was cowardly done and not by a Chevalier of honour and, if he had done it one-on-one, that he had not been able to do it at his ease, that he could not draw satisfaction from it. The satisfaction that he recognises his fault is sufficient, especially since he who admitted having done a cowardly deed is greatly dishonoured. For he cannot proceed chivalrously into a duel to make a proof with him who is not bold enough to be found face to face, who wants to kill him treacherously.

Chapter XXXI – Baton strikes and slaps[91]

I have never seen a thing to which Gentlemen and Chevaliers of honour have taken more displeasure, and which has been more odious to the nobility and to Chevaliers of honour, than to debate and decide their quarrels and differences with baton strikes. It is a thoroughly despicable act,[92] made against all the privileges of chivalry and about which the King must be a rigorous and severe judge. Such insults are made through the contempt that one has for him with whom one has a quarrel or by the pride and presumption that one has about oneself. What agony I pray you should an audacious and reckless person suffer who abandons all honour and respect to attack an honest Gentleman with baton strikes? This act truly seems to me to deserve no excuse, but on the contrary, must be avenged according to the status of those who have been outraged in this manner. A fool who has lost his mind must be excused but he who is sane in understanding and in full reason cannot be in any way [excused]. If this cause is called before the judges, they cannot condemn the culprit of a such a deed to death; only to some pecuniary fine and to some slight reparation. Still, it must use up a lot of time and consume his worth before being able to draw out any reparation. How should a Chevalier of honour thus draw satisfaction from the baton strike which will have been given him? I do not know of any other means than with his sword, that which he has with his life drawn satisfaction with it. Otherwise, if he does not do thus, he is dishonoured. For this reason, the King, when this quarrel reaches his knowledge, must so cleverly examine this villainous act, that he who received the injury of a baton strike be sufficiently satisfied for it, and must condemn the one who delivered the baton strike

---

[91] *soufflet*. The *Dictionnaire du Moyen Français* gives this as *Coup de main à plat sur la joue, gifle, soufflet*.
[92] *vn acte du tout abiect*

to be deprived of the honour of combat as long as he will live, enjoining him with never carrying a sword or other weapons, as one to whom it is not appropriate to ever carry arms because he despises the customs of the quarrel, which are proper to Chevaliers of honour, and also of never being able to call out a man to duel and to prohibit all others from calling him out because, having the sword at the side, he preferred rather to hit with the baton, and also to not leave his court and after one year or two or three or however long he likes, that he be compelled two or three times per week to present himself before his Majesty as testimony to his obedience, and to condemn him to follow the captain of his guard, who will be in the district in order that all the Gentlemen who ordinarily live in his court know his glories, and the boldness of it must be chastised when they forget, of giving baton strikes to Gentlemen and leaving their swords in the scabbard. The audacity and insolence of such a person must be reprimanded through humility. That can only be done by the Prince because he is as much dishonoured as he who thought to remove honour from his enemy. All Chevaliers who hit their enemy with a baton rather than with the sword perform an act which is indeed condemnable and against all privileges of nobility. We would not do any less to a valet. If this form is not found good, another may be found that his Majesty may command. But certainly the King must be a rigorous judge in this quarrel without exception or favour. The villainous acts between Gentlemen and Chevaliers of honour must be well examined and severely punished.

Second Part

Having already spoken of the duel that the Chevalier is bound to pursue in that which touches his honour and, having brought together all that which seems to me to be good for defining a duel, and also for when they are in the field, it seems to me I have declared all that is proper for the attacker and for the defender. I added also the contradiction from which I am of the opinion that most of the quarrels of today have proceeded. Have I done wrong? And to that end, so I can better examine from thread to thread that which is suitable for the quarrel, and so I do not make any confusion in this discourse, I thought that it would be good to do it in parts, separated one from the other. I will speak at this time in this second part of the challenges,[93] with some other discourse which could serve the subject.

Chapter One – About challenges in the King's Court[94]

The Kings' houses have always been respected and revered as sacred places, especially since the King is established by the will of God and has some similarity to the divine. Also, they are like lieutenants on earth, who have dominion over their subjects in order to be obeyed, honoured and revered according to God's commandment. In several places in his Epistles, St Paul calls them the images of God on Earth and, in many other places in Scripture, he expressly commanded us to honour, respect and revere them as much as is possible, even obeying their laws and ordinances although they sometimes seem to us more harsh[95] than gentle and fair. This is the reason why subjects must

---
[93] *appels* - callings out, challenges
[94] *logis du roy*
[95] *rigoureuses*

not insolently oppose their Prince, being something separated sufficiently from reason that members raise themselves against their chief and oppose his will. Nevertheless, it is something which is common enough today, [against] which the King must give a model order and chastise well those who would be so daring to call [someone] out to combat in his court without his express commandment. I am never more astonished as [when] the King is so patient to endure those who quarrel there and call each other to combat without [showing] any respect. That is punishable and the Captains of the guard should be careful, when such scandals happen, to arrest[96] them and the king to punish them. Again, I say that he who calls one out must be punished instead of the [one called out], although both deserve it. Yet, he who will call [someone] out on behalf of another only does it in consideration of the friendship which he holds for his friend and, without this, it doesn't touch him in anyway. If the other calls himself his enemy, it is from the animosity which incites and pushes to quarrel he who offended him. Nobility is acquired by virtue of our predecessors and by their prowess and valour and by their good conduct. It is necessary that, being called near the King's person, we are so well studied[97] that we are respectful in his principle residence and embrace well the virtue that we are recognised to have from an old lineage and good breeding. The gentleman can only be better known by these actions and demeanour. I saw in the time of King Francis I and King Henry II that we would not have dared to undertake to call out or insult each other. It was life and death to those who undertook it. Now, however, in the so blithe youth who haunt the Court, it cannot be avoided sometimes that quarrels arise for some badly said words or for the love and service of the ladies. For that, should another call out his companion, summoning [him] in the King's house, giving the lie, calling him out, or having him called out, it seems to me that it would be good to respond, "I only know what you tell me at this time and cannot comprehend your words. But, outside the King's residence, I will respond to all that you will say to me." And, at the same time, leave without otherwise calling out [the challenger], and if he with whom he had words does not leave, the other [the challenger] must hold himself to be in no way injured or offended. My opinion is that all the insults that are said in the King's house are made to the King and not to those to whom they are said. This is the reason why the King must take this cause in hand and make it a punishment, both for those who draw their swords in a company of guards or who are called upon to find

---

[96] Literally, to take them.
[97] *si bien apprins*

their swords and daggers in the view of the guardhouse in front of the King's house. All this is a capital [offence] and worthy of death. If the King had had someone made naked, torch in hand, accompanied by the executor of high justice, made three turns around the lower court of his château, and ask for pardon aloud to His Majesty, then, returning him disgraced to his house, you would then see the King's house well respected and his edicts well observed, and everyone would be afraid of committing such insolence in his house. Or else, condemn them to perpetual imprisonment. This is the punishment that I would desire for them.

Chapter II – Of the several types of challenges

Many types of challenges are practised today that have never been the custom or exercise of previous Chevaliers and, however many times His Majesty has opposed them with prohibitions, he is not always obeyed, so quick and fiery is the ardour and courage of the French gentleman. I will say and stay myself in this opinion that when a Chevalier is offended, be it by word or deed, he must be permitted to call out the one who injured him, but not by a *cartel*[98] or by a private friend. He must have the other called out before the King and ask permission of His Majesty, who must not refuse to allow them the duel, having first heard their differences, having also attempted to reconcile them and make them friends. By this, I conclude that no Chevalier whosoever be allowed to challenge his enemy without the King's will. I ask what assurance one can take in finding oneself in the place of combat on a simple *cartel*, however much he who carries it may be a true and honourable Chevalier who intends no deception. Nonetheless, why should we blame the relative or friend of the one who has been called out, knowing this, if he is found there assisting him or if he is called before the King, when His Majesty will regulate them in their duel? I laugh at those who have a friend call out a Chevalier, who speaks with him like this. "I have come to tell you that such and such a Chevalier that you have offended is near here with sword and dagger in order to seek satisfaction for the wrong that you have done him. And if you have a friend to accompany you there, let us talk together, him and me, for I do not want to put my

---

[98] *Cartel* is the technical term for challenging an opponent to a duel of hour by written message, sometimes published publicly, sometimes delivered privately by hand.

friend into a duel that I would not be party to." The other is ready and prepared to fight and not to lead another to it. In this, he is founded on good reason, but I believe that the one who has come in order to challenge has done wrong by the friend of the challenged, imposing on [him] to fight as a second, and that the challenged should not undertake it unless his friend consents [to be his second], and the one who has been challenged accepts the duel on this condition. I maintain that he is not called out and the calling out which has been given stands as nothing. And the Chevalier who has been offended remains offended still, and he must make to repair the offence which has been done to him by another means, and that he be chivalrous. But, if he is called out to combat one on one, he must accept it without exception, assuring himself that his enemy is so full of honour that he does not want to call him out in order to do him wrong,[99] and would not want to fight other than with similar weapons. This is the reason there are some who fight in shirts in order to prove the justness of their duel.[100] There are some that I greatly esteem who proceed there more chivalrously. When they have to ask for something that they feel offended by, they make the *cartel*[101] themselves and sign it and put it into the hand of their enemy, together assigning him the place of combat with full certainty [and] without advantage. This way is very laudable and, if he is not found [at the appointed time and place] where he is called out, he makes a great and irreparable error. I saw in my youth that we did not call each other out nor send *cartels* but we called out each other softly in the ear. We fought very often in this manner. I esteem and prize greatly this sentiment and this duel, for there is no braver calling out than that which is done oneself. This is the duty of a valiant and noble Chevalier. It is this calling out that all the valiant Chevaliers who want to emulate the honour of chivalry must follow, and I advise all those who may be offended to take this path. In this way, they acquire great reputation and honour. I will put in this place an example that I have seen. It was Sourdeval, a valorous gentleman who had a quarrel with a Captain, both were in the King's service at Fort Boullogne which the English held at that time. Their friends kept an eye on them and wanted to prevent them from fighting. They both conspired to feign being in agreement and good friends. This lasted more than six weeks so that the friends, seeing them talking together judged that they were in agreement, in the end believing that they no longer had to have regard for them. They secretly called each other

---

[99] *pout luy faire vne supercherie*
[100] *pour faire preuue de la seureté de leur combat*
[101] *billet*

out and fought outside the fort and were both heavily wounded. Their friends ran there and brought them back to their quarters. Thus, a brave calling out and two valorous Chevaliers and noble sentiment: there is no ceremony in this duel but that which is done dexterously, bravely and assured of courage. Also, it was boldly fought when no one knew about it.

Chapter III – Of those who send *cartels* of challenge by their servants

I believe that those who send a *cartel* by their servants do not take the advice of their intimate friends, and I think that all those who intend to decide quarrels never advise this, because sending a servant to carry a *cartel* is to scorn arms and denigrate the rank of Chevalier, which has been so honoured for all time that all valiant men have striven to attain the noble rank of a valorous Chevalier by such fine deeds as may be signalled and prized by all valiant men. Must a valiant man who has acquired great honour and reputation receive this affront that he is called out by a servant? And when he refuses to receive the *cartel*, he does the duty that an honourable Chevalier must do, being indignantly presented by someone unworthy of knowing so honourable an effect. Some want to send a gentleman in order to certify that the *cartel* that the servant carries contains the truth. If the gentleman accompanies the servant, I maintain that he is a spectator to the affront which the servant does to the valorous Chevalier, and I conclude that he must refuse his friend such a delegation, being shameful to accompany a servant for such a deed which must be executed more honourably. It is not honourable that a gentleman accompanies a servant who carries a *cartel* to an honourable Chevalier. The servants are only in the service of gentlemen to walk with their masters, to spur their boots and hold their horses when they descend and want to mount. I pray you, messieurs, to listen to the quarrels of honourable Chevaliers and how they conduct themselves. See if I am inappropriate and build reasons which may be evident in order to defeat my words. I will never change this opinion unless I am taught by some learned man who has better understanding than me,[102] because I always want to learn. I will say again at this time that you who depend on honour must always act honourably and that there is nothing there to take

---

[102] *qui soit mieux entendu que moy*

back.[103]

Chapter IIII – Of those who go to challenge the enemy of their friend in his house

There is a way to call out one's enemy in his house, for going there and knowing that he is not there, one must delay until he is back. And if he is very desirous to want to complete his duty, it is necessary that he make efforts to talk to the most important person from house, even his parents, if he finds them, and tell them the duty he has to their friend. In doing so, I believe that he has executed his duty and that challenge is well given. Let us suppose the case that the Baron de la Garde has sent a challenge to his enemy at his house, that he who went there did not find him, but found the enemy's brother instead, to whom he made it known that the Baron is ready to fight him for some words that he said to him inappropriately. Would you not want to judge this a good challenge? For when he tells the brother about the one who he intends to duel, the brother must make it known to his brother, as he who must participate for his honour and, having been made aware of it, must notify the Baron that he will not fail to be found in the place which he has assigned him. And, if he wants to allege that they have not spoken, notwithstanding, he still remains called out and his honour well engaged if he does not satisfy him. It is, as if in a criminal case, that one gives you a personal adjournment of three brief days and then, failing to appear, you would be in default. These two causes have some compatibility together, such that the one is in default by fault of not being represented on the assigned day and the other dishonoured by fault of not being found in the place to which his enemy has called him out. Thus, I conclude that all valiant Chevaliers, and those who are full of honour, will never allow such a calling out to pass without well satisfying it at the peril of their lives. One could say that these two causes have nothing in common together because the one debates before the judges of justice, the other with the sword. It is true but we advise responding to it. This same Baron de la Guard who is offended, injured and outraged, instead of calling out his enemy to combat, has given him a personal adjournment in order to repair the outrage which he has done him. If the enemy does not appear, he is condemned. Whereas, if he presents himself there, he is decreed a prisoner. Perhaps the outrage will be of such

---

[103] *qu'il n'y ait à reprendre*

consequence that the judges condemn him to lose his head[104] or to a reparation of honour which will be shameful to him. Thus, the two ways which present themselves to Chevalier. It is up to them to choose which is the most honourable and the most prized. I think that if the enemy of the Baron de la Guard is old and very decrepit, even powerless to handle weapons, he must be excused. But if he is healthy, strong and robust, he must satisfy him with the sword and, if he does otherwise, he will be held as a coward and lacking courage. Chevaliers are obliged to obey the law of honour which holds that when someone is called out into proofs of arms, he must promptly prepare himself with bold courage, and he who does otherwise is not worthy of being placed in the rank of an honourable Chevalier. I speak after the fashion of men who want to follow the customs of valiant Chevaliers but not divinely, for God has reserved vengeance for himself, as being the God of arms and duels, and gives victory to whom he please as being a very equitable judge.

Chapter V – Of the one who will call out his friend's enemy in a house other than his own

I will say about this calling out that the lord of the house must say to him that it is not appropriate to deliver the duel [in this way], for delivering the duel on his own authority is a crime against the Crown,[105] where he only does it with this in mind, and for this reason the calling out must stand as nothing and without result. Because it is not legitimate, I believe that he who sends to call out his enemy in the house of another, whatever excuse that he makes, does not warrant coming to blows and I think by this means we will put them into terms of agreement. This is one reason which is strongly apparent, and yet it is not chivalrous and, if it is necessary to tell the truth to him who makes such callings out, making this pitiful office makes his life cheap. Also, this act cannot be said to be chivalrous but is understood to be executed by a rash man full of arrogance. I will say, moreover, that the lord of the house and the one whom he wanted to call out must not respond. This calling out stands as nothing. The lord of the house could say, "Tell your friend that if he does not better satisfy the calling out for which he has sent you here, that I hold him a coward, because he could not find his enemy elsewhere than in my house. And to you, no longer

---

[104] *à laisser la tete*, lit: to leave or allow the head
[105] *c'est crime de leze Majesté*

make such affronts to a gentleman of honour, because I could chastise you strongly for the wrong that you have done, if I wanted. But because I have the advantage over you, be satisfied with this rebuke so that I remain content." For as the one carrying the challenge has done displeasure to the light-heartedness[106] of the lord of the house, this lord can do the same to him by not accepting the challenge.[107] Because he undertook this calling out as a bad Chevalier, also must he be chastised for the wrong that he did indiscreetly.

Chapter VI – Of those who present themselves to second their companion

I want very much that those who ask their companion to second them in combat, or those who present themselves for battle two-on-two or three-on-three, to give me apt and evident reasons why they do it. It's a thing which has never been seen or practised except (as I said at the beginning of this treatise) in legitimate war. I will remain, therefore, of this opinion that in order to end a quarrel, it is not necessary to call a second to it. He who is offered combat in pairs[108] or who asks his companion to assist him in person, nor any less he who wants to suffer it, does great damage to his reputation. There are several plainly evident reasons to prove this form of duelling wrong. He who seconds his companion has no quarrel with the one his companion fights, and perhaps he is the best, most familiar and private of his friends. Thus, the great cruelty of fighting in this manner and without a quarrel. It is too barbaric to force this choice on a friendship. There is more than the one of the two who will be affected by his enemy that, if he sees his companion too impeded, he will try to rescue him. In doing so, he carries out an assassination and in this fight there is no honour. There is another reason that we could suspect: that most of those who want to be seconded do so to have greater assurance and, seeing themselves thus accompanied, they are more resolute, because one would like to say that one fights more determinedly and with more resolution being accompanied than when one is one-on-one. Also, I have always esteemed a valiant man, who secretly and without it being known,[109] pulls his enemy by the cape into one on one combat. In this calling out, there is a lot of honour. I nonetheless believe that one does it for better

---

[106] *il luy a fait vn desplaisir de gayeté de coeur*
[107] *ce seigneur luy en peut faire le semblable sans acception*
[108] *combat en deux*
[109] *sans le sceu de personnes*

reasons than I can think of. If the duel is fought with a second who has not practised quarrels for twenty or twenty-five years, misfortune will come of it. I remember on this point having heard about the late M. Francis de Guise of Lorraine and the late M. Sensac, the father lately dead, valiant men, and their most noteworthy deeds and their words, the said lord of Sensac considered very brave and bold the one who goes thirty of forty steps in front of his companions or the one who will give a spear thrust the day of the duel. Monsieur de Guise replied: "I beg you, Monsieur de Sensac, do not put that in your opinion and believe that the beautiful and great companion is indeed the cause of these boastful ones" – as he called them – "who rush to do similar fights and do not think that there is boldness when it is done out of envy or jealousy, but take for a bold man the one who calls himself secretly to fight his enemy. These really are valiant who know well how to fight their quarrel, without anybody else getting involved with them, and even more honourable when we know nothing about their fight until after it is over and named, for example, as some who so fought." I wanted to recite his words because they come from the mouth of a great person and one who is greatly recommended and who has acquired the first rank among the great Captains, which serves well as witness to certify my words.

Chapter VII – If one must call out his enemy at the head of a company

This request deserves to be properly resolved and I am of this opinion: it must be permitted for the gentleman or the soldier whose honour is offended to call out his enemy at the head of the company, and the Captain cannot complain about this, even though he marches with the ensign unfolded. But if it happened that he called him out and picked him out of the rank where he was to have his satisfaction, he would be making a great mistake, and then the Captain would have reason to complain and interrupt. But, there is a lot of honour in asking the Captain's permission and he cannot refuse it. However, he can put this quarrel before his camp master and the Colonel to deliberate and give him their opinion, according to the advice and counsel that will be taken, because quarrels must be ended in front of the superiors and those who have authority and power in the army. It is not allowed (according to my judgement) for a simple Captain of a company of foot soldiers, or of a company of men-at-arms, to fight the duel unless he has a superior [present] who commands him.

It is he who has the authority and the power to deliberate disputes and duels. But if there was not yet an army corps made up and the company was marching alone, the Captain must bring the soldier out of the ranks, and do right by the one who asked for it. But, where there is a Lieutenant-General of the army, the power of arms is given to him, and not to the simple Captain. Polybius says that the consuls formerly had power of life and death over men-at-arms without any appeal and, for this reason, he said that they had royal power. Also, the power of the sword is given and reserved to the Constables and to the Marshals of France in the conduct of the army. It is quite reasonable that the King, in order to avoid the scandal, provides in the military ordinances of his Kingdom, even to Captains of *gendarmes* and foot soldiers, to regulate their soldiers and oblige them to follow and hold the laws and customs of war and the form and manner of the military art which will be given to them without adding to or removing from the [laws], on pain of death.

Chapter VIII – If the yeoman should call out the gentleman to duel

It's necessary to talk of this calling out, which is a discourse worthy of being well understood, in order to give justice to him to whom it belongs. Before speaking of this calling out, it is necessary to know how it must be received between Chevaliers. We will speak of nobility and, as from ancient times it has been esteemed and greatly prized in those who have acquired this grade of honour and, having acquired it, they have managed to preserve it. This is the path those who want to hold the rank of nobility must follow. Also, it is necessary that, in exercising it, he do no kind of disgrace, otherwise he would make a stain which would turn it to his dishonour and which could be a taint on him and his descendants.[110] This nobility, which is preserved from time immemorial with much honour and honourable reputation, and which is maintained in honourable duties for the service of the King and of the nation, having risked by it his goods and his life, must be prized by everyone and recognised not only by the aristocrat[111] but also by the yeoman, who must respect and honour him as coming from an illustrious family and from another house and other parentage than his own without making any comparison with it. Otherwise,

---

[110] *posterité*
[111] lit: *noble*. The overuse of the word *noble* in a variety of senses becomes very confusing here.

if this took place, we would all be equal and alike. This must not and cannot be. Yet, today, the yeoman, for the short time that he has carried arms, calls himself a soldier. And how often is it with only three or four months of carrying the sword at his side and the arquebus on the shoulder that he wants us to hold him a notable soldier. He does more. He will line up with the swashbucklers.[112] He shapes himself to draw the weapons, after a fashion,[113] which his father acquired with good means, means which are good enough to be held in honest company and with a fine ostentation in order to be found with gentlemen where he is welcome. Thus, being put into the rank of the nobility, he wants to be maintained there without exception. He tells himself he is, and has himself proclaimed, a man of quality and honour and has others say that he is such, and that he carries a good sword to maintain this against those who would doubt it. Thus, from beautiful and very proud words, it only remains to enact them. He takes a quarrel with a gentleman of honour and a strong man of means and has him called out. I ask how this calling out must be handled. My opinion is this: that the gentleman must respond to the one who called him out as a soldier, saying that he does not march to his orders, not that he dreads and fears his sword, or his valour, but because he recognises and knows well what he is, that they can be seen together and thus he will give him justice with all satisfaction he wants from him. I will say this word in passing that I counsel all gentlemen to not enter into proofs of arms other than with their equals and to not match their sword or their courage with one who is not of their kind and who cannot be compared with them. The misfortunes which could happen are very dangerous. It is very reasonable to avoid them. But I will say that if this soldier was so strongly offended, he could himself go to find him in order to draw the sword, when the gentleman cannot not produce any good reason for why he did it to him. Yet saying to the soldier that they are not similar, it would not be speaking like a noble man. He must fight at whatever risk may happen. But if the soldier, before coming to arms, found the means of obtaining some honest satisfaction from him, he would be wise, and the gentleman should do so, both for the honour he bears in arms and for the desire he has not to offend anyone, without nevertheless doing it on his own. For such people, when they are so glorious and so insolent as to have drawn the sword to come and confront them, deserve some chastisement. It is not reasonable to endure such insolences, and those who support them in these actions have done the greatest wrong. I

---

[112] *espadassins*
[113] *tellement quellement*

do not speak without purpose because I know those who preserve and debate all their rights without regard to gentleman or yeoman, and make them alike. I marvel that the gentleman has no respect for nobility, which is something that must touch him very closely. It does not only offend this gentleman but all those to whom he belongs by blood. What I say about the yeoman is not that I want to put up a barrier to prevent the yeoman from being able to aspire to nobility, because nobility is acquired by virtue which can be exercised both by the yeoman and the aristocrat. If the yeoman is ennobled, it is understood that he has it, for him and his family, because it could be that he had poor parents who could not have followed this vocation. And yet, in order to be taken for and esteemed as a gentleman, it is necessary that this comes from father to son to the fourth of fifth generation, and still he could not be called a gentleman of name and arms and of an old house if their quality was debated. But because he chose arms for his principle objective and exercise, it remains with his family to always maintain themselves by acts which closest approach virtue. He is ennobled, having acquired honour and reputation such as a brave soldier must seek, in order to be esteemed and held to the ranks of the nobility.

Chapter IX – Of close relatives who call out each other to duel

It is a shameful thing when near relatives quarrel. When friends and neighbours and close relatives know of a quarrel between them, they should try to agree, so that this quarrel does not grow in size. Certainly there is no enmity so great as that between relatives when it has rooted, and uneasily can it be pacified when there remains some discontent, and the worst is that the quarrel very often is so great that they are forced to call each other into duel, each for their share [of the inheritance], which is the most frequent quarrel that occurs between brothers, where it is necessary that the arbiters, granting the civil and the criminal, have the almost impossible task of bringing them to agreement, so obstinate are they against each other. In the end, only blows and sometimes murders come out from it, which are found mixed in these histories. In this, the house is dishonoured and despised for seeing [its members] so cruel enemies. Quite often there arise other disputes between the brothers which proceed on the occasion of their wives, I am ashamed to say it. But it is of such consequence that they are called to battle, and do not see a sole means, the subject being so bad that the honour of a good man depends on it, and that whatever fraternity there may be, that it must be fought with arms. It is a great cruelty and a thing totally inhuman that he who ought to be desirous of honour of his brother's house, is he who dishonours it. He is shameful to the brother entering into a dispute with him, as even the youngest who must by right of nature yield to his eldest brother and be careful as long as it is possible for him not to offend his friendship but must keep it dearly. When I speak of brothers, I intend also to talk of the uncle, nephew and close cousins,[114] brothers-in-law and other close relatives. And when there is a lawsuit between the brothers and close relatives, this dispute must be referred to arbitration. Because the brother will always have more honour towards his brother when they share the proceeds with an honest affection than to fight with full severity. It is sometimes necessary to yield to greed and to maintain themselves in friendship and harmony. I will produce a proper example for the truth of my words. Athenodorus had a brother who was his elder, who lost all his property by justice. Seeing him poor and destitute of means, Athenodorus shared the rest of his property and gave him again half of it. The wise gentleman will always be careful not to argue with his brother, uncle, nephew and cousin because from these depends the support of him, his children and his house. Because when a family maintains itself in good love and harmony, it is much more praised. By this, they make known that they have come from an ancient and illustrious family which does not want to offend the friendship of even one of their relatives but wants to maintain all gentleness, peace and unity in a way that, a quarrel touching one, touches the others. Such alliances are to be feared and should we seek them.

Chapter X – Of two combatants, if the one who retreats must be accused of cowardice

It is the common opinion that he who retreats in a duel is maintained to not be daring and bold. I am not of this opinion and believe that every man who draws a sword when he is called out and parries blows and defends himself must be a valiant man even when he retreats because between fleeing and retreating there is a difference. To flee is to go and retire without wanting to fight. To retreat and parry blows is to defend oneself and to await the chance to defeat his enemy, for [he

---

[114] *des cousins germains*

who makes] such a retreat afterwards approach bravely. I intend to talk about one-on-one combattants. But if it happens that one alone is assaulted by two or by several then it is no shame to withdraw and retreat, even to run,[115] and he who was thus assaulted could say, "You are too many on one but if the most valiant of you may come one-on-one, I will fight." And in saying these words his honour is in its entirety. Therefore, he who retreats in single combat is not a coward, and I say with advantage that he who fights furiously and advances himself more than the other, and throws more blows at him, he should not be considered as the bolder, and I believe that he has no more judgement and boldness than the other who knows how to run and strike when he sees the opportunity. This latter is a brave Chevalier, resolved and very determined, of whom one could well say that he drew the sword despite always retreating and not throwing a single strike. We will respond to this question [by saying] that is it enough that he is well covered and prevents his enemy from having the power to injure him. It is a lot in combat to know how to be covered and not be injured. Let us speak at this time of he who in combat finds his enemy alone[116] and throws himself,[117] with his sword and dagger, at him and wounds him, if we should say he is well fought. I will respond to him if he flees and if he turns his back, he is well fought, but having the sword in hand defending himself, he who would have done such an act must be judged as having done wrong, because to throw oneself with the sword or the dagger is as much as if one changed arms and took a halberd to strike from further away. This is to be judged and condemn oneself as not being strong or courageous enough to fight one's enemy, since one is afraid of wanting to fight with such weapons.

Chapter XI – If one of the two who have a quarrel convinces the other to draw a sword, even though he doesn't want to do it, and the first wounds the second, it is badly done

The Chevalier has always been held as generous and valiant when he had a quarrel and did not take his enemy at his advantage but finding

---

[115] *lors ce n'est point de honte de s'en aller & retirer, voire au grand pas*
[116] This seems in context to mean unarmed, or perhaps armed with sword alone.
[117] The original uses the verb *darder*, to launch, throw or to wound by piercing. Context trumps all.

him one-on-one made him draw his sword or had him called out. And if it happened that his enemy does not want to defend himself and asks him, "Do you have something to unravel[118] with me? You have spoken ill of me and have done me an injury." His enemy told him that he had never spoken of it. I believe that he must remain content and well satisfied, because he is having the injury repaired that he had done to him, having the sword drawn and one-on-one seems to me that the denier must suffer it. Lord Muzio is not of this opinion and concludes in order to deny it that the Chevalier is not satisfied and that he remains still offended. I hold the opposite, that not wanting to take advantage of his enemy having the sword in hand and one-on-one he has been shamed and done wrong to his honour, and the Chevalier has acquired as much honour and reputation without his enemy making him greater reparation. But if it were that he did not want to draw his sword or to satisfy him, I believe that he would not have done wrong to wound him because it is the true testimony to have drawn his justice from the wrong that his enemy would have done him, having arms in his hand and without advantage, which are very honourable acts, and that all Chevaliers must take for a coward the one who did not want to draw the sword to make justice with his enemy.

Chapter XII – Of two who are in a one-on-one duel and one seizes the sword of his companion then hits him, whether he has fought well

Many would be of this opinion that, two Chevaliers being in combat, the one who would seize the sword of his opponent and wounds him would be well fought, or if the opponent were killed, that the survivor must not be badly judged. I would be of the contrary opinion and could not approve the valour of the one who had fought in this way and would judge the combat to be the same as if he took his enemy unarmed. I know very well that there are some who would counter me and say they do not fault their honour but I pray the one who would hold this opinion to give me one reason why he seized his enemy's sword. I believe that he will respond to me that it is to deprive him of the means of offending him and having victory over him. It

---
[118] *demesler*

is the most obvious that he can give me. I do not at all approve of it, especially since the Chevalier who has debated some quarrel with his enemy, must debate it with the sword, because by that we know the valour and the boldness of each one to take the sword of his enemy. I do not find that there is boldness in it, but only rushing and fighting in desperation for the fear that one has of one's enemy. I would argue that if the sword of one of them broke, or that it was pulled out of his fist, and that he was so strongly charged that he could not recover it, when he seized the sword of his enemy and threw himself at him with the dagger, he would have valiantly fought to make him abandon the sword without however wounding him. If in fighting, one happens to close and join with his enemy, wounding him with the dagger and throwing the sword to the ground, knowing would it be well fought? I answer that it is boldly fought because it is done at the risk and hazard of life. But I would be of this opinion that he who would have done such an act must for this must not have his honour rebuked or be faulted for valour because sometimes the disposition of one is so often the cause of such an event. But this would be much worse if, having thrown the sword to the ground, he removed his weapons from him. One can accept beating from the enemy's sword with the hand and not [accept] taking it.

Chapter XIII – If someone pushes you rudely from playfulness,[119] must you call him out?

When someone carries bad will[120] for another passing freely near him, he pushes him rudely. In doing this, he seems to want to provoke him and quarrel with him. This is done without purpose with very little consideration because if you have something to ask, it is much more honourable to ask it with your mouth than with a knock. And if it is up to him to do so much more you have an advantage over him for it is a sign that he does not dare to do it, or that he fears you. We know well enough the quarrel of Messer Carensi and the Baron of Biron, he who had been Marshal of France. Many honourable people know how this quarrel ended. There were others who with playfulness knocked and pushed each other very rudely without having any debate but often they were not known. I remember having granted one a quarrel for a similar subject for two nice men who had not seen

---

[119] *gayeté de coeur*
[120] *mauvaise affection*

One-on-One Combat in the Arena

or known each other. This quarrel was put before Monsieur le Constable, where both being assembled there, the elder of the two gentlemen begged him very urgently to have granted by the King the duel against the one who had offended him so inappropriately, not knowing him and not having given him any cause. In order to decide this quarrel, Monsieur Admiral Chastillon, the Marshal de Vieille Ville and the Grand Squire of Boisy[121] with several other lords were called there and for the excuse, he said, "What I did was not to give you any offense or affront or that you had given me any cause to do so. But I did it taking you for another who is accustomed to frequent a gentleman who is my enemy, and taking you for that one I did what happened between the two of us, about which I'm very grieved. I beg you to excuse me, and that we remain friends." Here is the apology[122] which was made to him, with several challenges on the one hand and on the other. To say my advice: I would argue that when someone has playfully pushed another, that at the same time and that on the spot he takes him by the arm and stops him to find out from him for what cause he pushed him and that he wants to be satisfied, and still that if was in place of respect, he must not delay, instead, saying to him, "Let's go out of this place," because to delay five or six hours or a day to call out your enemy after being pushed, I cannot approve of. It is a calling out seen when there was proper time to draw his reasons for it without delaying things. This cannot be turned back on the reputation of he who delayed thus his calling out and if he wants to excuses himself by saying when he had been so rudely pushed that he was in a place of respect, I answer that there is no place of such great respect that at the same time one cannot find the means to leave there without reflecting on one's call. To push someone playfully is to wrong him in such a manner that I would think when someone feels himself rudely and with playfulness pushed, if he would take the collar of the one who had done him this affront, with a dagger, he should have satisfied his duty being nonetheless out of respect.

---

[121] Could also be rendered as the Master of Horse or Cavalry Master
[122] *satisfaction*

Chapter XIIII – If a gentleman in service must be considered equal in combat to a gentleman of honour and from a good house

Servitude has always been assessed that those who have submitted to want to serve gentlemen have been held to be domestic servants, so that being reduced to this point, they must be at all times prepared to obey at the command of their master, which must be called servile obedience. This obedience, it must be explained is when a gentleman servilely begins to obey, and when he is in the wages of another gentleman and serves him and makes a pledge to him, that one being in this state must be considered as a domestic servant, who is fed, dressed and maintained by his master. That one must obey all the commands that his master will give him, and cannot measure himself against another gentleman who is of good stock, quality and good house to have satisfaction for a quarrel which would be caused to him, and cannot call him out. For since he is a servant in the pay of another gentleman, that lessens the other to him that he must hold for debating his quarrel. I know well that there are many gentlemen who have gentlemen in their service, who will not be of my opinion, and would like if a gentleman had a word and dispute with any of their gentlemen that he could have him called out. But I can only consent that an appeal is made by a gentleman to whom he is pledged, because I consider him a servitor, not that I mean to say that he should lose the title of nobility, nor that he should be outraged, but because is among the ranks of servitors who cannot be equalled to a gentleman of quality, honour and house, and who, given the state he is in, must be it is that he yields and use the respect as a servitor. I do not want to conclude that a gentleman is dishonoured by serving. It is better for a gentleman to put himself into service than to do a worse office and lead an unhappy life to endure after a shame which would reproach him and his family. Between similar, however, they can be called out, after granting them their differences, but not with one greater, if it was only his master wanting to take up the quarrel. As being aggrieved that one would insult his servants, we could for love and contemplation of the master and his friendship say a word to him for his contentment, but not as by way of an accord or form of reparation. This is why the gentleman who puts himself in service must think well before putting himself into this bondage. Firstly, he must represent that poverty forces him to obey, and that for this reason he resolves to obey well. Secondly, that he intends to be very respectful to close intimate relatives, friends of his master, and to take good care not to do or say anything that he might receive any displeasure. For that would make so many enemies for his master. When he is in service, he must comport himself as a servant.

Chapter XV – If a sexagenarian gentleman must be exempt from the duel

It is very unhealthy for an old gentleman to be quarrelsome and should be in control of quarrels, having reached the age of sixty years, which is the age that all must more gently and wisely conduct their actions but also it could happen that he would have such a debate in which his honour would be engaged to which the man of honour and valour cannot yield, and for his duty he must draw its cause from it, although he was aged sixty years, that all maintain the Chevalier to be exempt from combat. However, I hold that if he is healthy as there are some who are as strong and robust as one who is only forty years old than he must not be exempt from combat, if he is called out, or if he is offended to call out his enemy. There are old gentlemen who are so courageous and have been such valiant men, having acquired such an honourable reputation who do not recognise even one kind of cowardice and are so healthy and fit that they will valiantly resist all who will want to attack, without wanting to excuse themselves. King Francis I granted a duel at Moulins to two gentlemen, one named Lord Veniers and the other Lord Serzay, of whom one was more than 60 years old and both fought valiantly, the eldest was highly esteemed. It is a rare thing to see a Chevalier of sixty years going into a duel.[123] I know of no other reasons that honour and valour to command doing many things with weapons. And a valiant old man knowing himself healthy and vigorous cannot be content to be offended that he doesn't take up arms. There was a quarrel in Paris between the young Chateauneuf de Bretagne, and Chainay Laille, already very old, for whatever word there was between them. The young Chateauneuf, resenting this, sent for Chainay Laille, a very brave and valiant gentleman in the Isle of Antragues. Chainay was killed there, and young Chateauneuf wounded. This is how this fight ended between the young and the old, it is to prove the value and the boldness of an old gentleman. There

---

[123] *passez entrer en combat au camp clos*

have been very great Captains who have acquired the fame of valiant men who were older than sixty years. It turns out that Massinissa, King of Numidia, lived until the age of ninety years. One year before dying he gave a battle that he won. The next day he was found in front of his pavilion, bare head eating bread black as he was of strong and robust constitution.[124] It must be concluded that the old man, even though he is sixty years old, should not be exempt from combat when he is called out and if his honour is offended by it he must take revenge by arms. And if he refuses the duel, he is obliged to give the reasons which may be legitimate and to his honour. Otherwise, he has wronged himself. This is an opinion which will not be received by all. But also, I judge that to will not be rejected by brave Chevaliers who know that it is necessary to repulse insults and, because a gentleman has nothing so dear as his honour, he must very honourably preserve it.

Chapter XVI – Of him who speaks on behalf of his friend

This is the question to know if he who is employed by his friend to carry some message,[125] must they conduct it? This question is diverse and because we have spoken above of the *cartel*, i do not intend to speak of it in this chapter, because it was a *cartel* which was for a duel, but in this one [chapter] here it must be understood in this manner: if my friend asks me to carry word to him with whom he has dealings, be it litigation or division or arbitration, I must freely concede to him, especially as the office of a good friend is to entreat a peace and accord between those who have a difference. But if he asks me to carry words of insult or other language, which may not be appropriate for Chevaliers of honour, it would be wrong for me to employ myself for this object because it is wrong for a Chevalier to carry words which offend the honour of another. The Chevalier is well advised to never undertake to do such an act. It is therefore necessary that the friend be so important to reserve his friend for a matter which is more important, and not to ask him to speak words which offend a Chevalier of Honour. However, in order not to reject at all the opinion that one would have of speaking for one's friend, I would greatly consider that it was practised in this way: I am in doubt whether someone has

---

[124] complection

[125] *pour porter quelque parolle*. Note: a *porte-parolle* is a spokesperson

said bad words about me and wishing to be clear about it, I ask one or two of my friends to go to the one who said them, and say to him, "We are here to know if you have said such and such words." I would greatly esteem the request [made] in this way and in this there is only honour, and the party could not complain, and according to the reply that he will give to him, he must withdraw without speaking to him further and let it be known to the one who sent him. There are some who send letters of injury by their lackey who are most often well beaten and sent back with the letters. It is also a thing in which there is neither honour nor any decorum because it is folly to have others say what one must say oneself, and mainly by men of vile and abject condition.

Chapter XVII – How this word "to feel" must be understood[126]

If the Chevalier who has acquired a reputation is offended by someone in his honour, he must not let this injury pass without resenting it and making proof of his generosity. This is the most common way of doing things by Chevaliers. Nevertheless, this word *"to feel"* is taken in several ways: when one feels like having come out of an honourable house, related to illustrious personages, and generous fathers, and when one desires to imitate one's predecessors in all virtue. It is to feel oneself and to recognize oneself for not wanting to do an act that degenerates from the virtue of these ancestors. This is properly called to feel oneself from the place from which one has come out. These are imitators of all honour and virtue, and "this feeling", is taken in good part, there is another resentment, even when it is a question of a quarrel because to speak properly of this term it is more common and much more in use for the quarrel than for something that one could speak: because when someone is offended he willingly uses this term, I will feel it, which is worth as much to say as if one said I will draw my reason from it, and will not spare my life there: and in fact every man who is outraged, even the Chevalier who must have his honour in more singular recommendation than another who would not be of such order, and if he does not resent the insult that has been done to him, he is taken for a Chevalier who is worthless and who has no resentment. We have quite commendable examples of our time, among others from the lord, d'Allaigres, son of Monsieur de Milland,

---

[126] *de ressentir*

I believe that I would not be out of place to recite it in this place. The said lord, feeling the death of his father whom the Baron de Viteaux had killed, knowing that he could not be right by justice, called the valiant and bold Baron, and who had repeatedly proved it and fought with the sword and dagger, being only twenty-two years old: an act certainly generous and executed with a bold and valiant courage, it was a resentment worthy of a son who, wanting to avenge the death of his father, still risked his life in his tender youth, a laudable thing: this term to feel this takes a bad part here, because it is in a quarrel that these resentments are made, and one can only very uncomfortably settle a quarrel there and feel an insult without being called and putting the sword in the fist, not that I want to say that the act is malicious, and entirely reproachable: but because it is done with rigour and anger, I give him this name, that is to say one feels the sword in the hand, and the other resentment of which I have spoken above is done in an honest way. The courageous Chevalier always feels the displeasure which he has received, so an honest man feels willingly the pleasure which one of his friends has done him, that which will do not another who will not have and will not know any civility.

Chapter XVIII – How the Chevalier must feel when he is offended

It has been said above said how this word to *feel* must be understood. It is necessary at this time to speak about how the Chevalier must feel of an injury who has been done to him. The Chevalier therefore who is offended, if the offense has been done honourably to him, the resentment must also be done honourably. For example: if someone has insulted you one on one and you answer him not at the time but some time later and from a distance, or through a window, or write to him that he lied, this resentment is not honourable in that the manner of the offense cannot be said to be well and legitimately satisfied with the offense that has been done to him. Example: if you are given the lie peer-to-peer, and you repulse this insult with a baton strike, you are insulting the one to whom you are giving the beating, and for that you are not satisfied by the denial which one gives you and remain always in dishonoured infamy, and not the one whom you will have so loosely and villainly insulted. The Chevalier therefore must be so advised that when he wants to give to someone a word of insult, he must stand firm to show that he wants to maintain his words and not to flee without waiting for the response of his enemy.

Similarly, the one who is outraged by someone, if he answers him with a denial, he must stand firm after giving it to show that he is a Chevalier of Valour to maintain him and force his enemy to resent it. This is how the resentment must be honourable or dishonourable. For, if you are offended by another you will have him called out or will command him to some place where you will find. You will make him draw the sword or make him hold the letter of the field in order to be found in the lists with the Prince's permission in order to debate your quarrel. In order to properly understand honourable resentment, it is when equal in arms and company, one puts the hand to the sword, or one-on-one or two-on-two one makes a proof of his courage and valour. It if this resentment that I desire the Chevalier of honour to conduct himself and following this path he will never stumble into dishonour, and if otherwise he is resented with dishonourable and illicit acts, he will acquire the name of an infamous villain and of a Chevalier who has lost his honour who can no longer be put on a rank in order to fight the injury which will have been done to him. It happens sometimes that two Chevaliers taking a debate before the Prince, of whom one is injured. He which is injured must not fear to give the lie before the Prince. For as the one has had the respect to have his adversary injured in the presence of the Prince, the other is no longer obliged and the Prince must not be aggrieved in excusing them honestly and honourably and with honourable respect. I am of this opinion that he who is offended must be more supported and favoured than he who has done the outrage.

Chapter XIX – Of the refusal many make in their quarrels and of asking their enemy's pardon

There are found very few men who want to demand their enemies' pardon for some offence that they may have done them and also very few arbiters who counsel their parties to want to do it. Cicero in his oration which he made to Caesar in order to have his pardon[127] on behalf of Ligario used these terms: "I said to Caesar, often pleading with you before the judges, but I have never said for he whom I defend, pardon him, messieurs, he has erred[128], he did not think of it, it is to the father to whom one asks pardon." Cato the Younger,

---

[127] *sa graçe*
[128] *il a failly*

being besieged in the city of Uticqa by Caesar and reduced to extremity, [was?] advised to send to Caesar to talk about an agreement and putting himself at his mercy. Cato replied that one goes to him there and that one does not speak of him, saying that it was for those who were defeated to pray, and to those who had done something great [it is] shameful to ask for pardon. There are sometimes injuries so indignantly made and against all [rights/duties?] of chivalry that I think that he who has attained it is happy to come out of it to ask for pardon. The one will say, but to ask for pardon is to make an honourable amends. The other will say that it is dishonour going to a gentleman to force him to a pardon. On this diversity the arbitrators will advise satisfying honest men without asking for forgiveness, and will say pulingly this word *pardon*, and let us transfer it to another that is easier. Some want that we say, excuse me. Others say, put it back on me, as being the term more appropriate for agreements. On all these opinions, I conclude that when for an agreement it is necessary to ask for forgiveness, that he who must give it must not fear to do so, because to say put it back to me instead of forgive me, I only know that is to say put it back to me and say that this word to put it back, is not clean for quarrels nor for agreements. There are those who wrote of combat but they do not take it for satisfaction and they do not say that putting back equals a pardon. I will talk of the remission more amply after this chapter and also of satisfactions, when thus one should make an agreement, and that one of the parties has greatly offended the other's honour, he must not fear ask his pardon, and the arbiters on the one side and on the other must hold hands there. Yet for an injury made to a Chevalier for whatever reason that may be, (when it is manifest and carries dishonour to the Chevalier) it is usual to ask pardon. And saying that if he asks pardon that he is dishonoured, I answer that the other is also dishonoured being outraged. By which I conclude if he must be dishonoured by it that one is dishonoured by the deed, the other by the words, and that they must also agree, there are also those who in their agreement want the [other] one to apologise.[129] I will say that I would like as much [both?] to ask pardon as to apologise, if this was only that one had given the lie to someone to his advantage, and that his enemy was put into an obligation to state his reasons at the time and that had been trapped in it. In this case, he who had advantageously given the lie must submit to such reason that praying him to apologise, with some honest words that the arbiters and mediators could find. But with gentleman to gentleman,

---

[129] *qui en leur accord veulent que l'on leur demande vne excuse*

he should not apologise,[130] according to the quality of that which I have said. Apologies must be made to the King, to the Princes or to whatever great lord. It is the function of any man of bad judgement and who has little understanding to be excused.

Chapter XX – Of the pardon that some ask for the satisfaction of their offences

Many are in this error of saying to their party, "I beg you to forgive me the offense that I have done to you." It is to well satisfy his party and just as if one asked for pardon. I cannot approve of it because I find that in matters of the quarrel that giving back[131] is much more than to pardon. The one mean that to put back is as much as to pardon. But that the one is milder, there are some who have written this remission, but they do not take it for pardon and say that for great injuries, there are some who want to freely put themselves in their hands and at their discretion, and think that such a reconciliation is neither good nor honest to make, seeing that if he who is offended takes satisfaction with his hands, he does not do it very courteously, and that such reconciliations most often one sees the quarrels start again, if it had been thus granted between the parties, which I have never seen done and think that there are few men who would want to do nor want to advise doing this, because to put oneself back in the power of someone who is offended in order to get his reason from it. It would be much more grievous than asking for forgiveness. I say further that if the quarrel is cooled and that it may be dealt with by an injunction, it is more appropriate to treat by remission. I have seen some who are so stubborn and have been firm that remission equalled forgiveness. I am sure that they change this opinion into one that is more suitable for quarrels and satisfactions. And in order to confirm my reasoning: forgiveness is a divine thing and it belongs to God alone to forgive our faults. He reserved it to himself when we ask him for forgiveness. Men must pardon each other, one to the other, when there is dissension between them. It is the commandment of God. We should not usurp that which belongs to God. This would be abusing his divinity because among the gifts of God, forgiveness holds the greatest and highest place. For our forgiveness, the son of God came in order to wash for once the sins of this world. Thus, as through lack of well

---

[130] *il ne faut vser d'excuses* - he should not make excuses
[131] *remettre*

understanding that it is only forgiveness, the arbiters who are called for quarrels are mistaken thinking that forgiveness must be less than pardon. There are term which are more proper for satisfaction.

Chapter XXI – Of those who do not want to confess the cause and subject of their quarrel

All the satisfactions that are made on the insults must be based on the truth, so much so that the one who is wrong must avow it. Any Chevalier therefore who charges his companion with an insult is bound to repair it for him, because the Chevalier who has his honour in commendation would not want to suffer that his reputation and his honour would be diminished, and is desirous of preserving it, wanting to expose his life to it: I will therefore say that ceasing the subject and the effect of the quarrel, there is no way of being able to honestly [bring to] agree the two contenders. No one can deny that in such an enterprise, the truth must be followed: why the Chevalier must be careful not to disavow what he does or says: also the arbitrators must be careful not to proceed in a quarrel the subject of the quarrel is not exactly well admitted, and then afterwards they will proceed to it much more dexterously and to the satisfaction of the parties and according to their desire. I know that in some agreements the arbitrators want he who gave the injury denies it completely or that he made no mention of it, of which I cannot approve. Because since the word is spoken, we cannot accept revoking it or even denying it, but one should repair it as honestly as possible.

Chapter XXII – On satisfaction

There are satisfactions made between Chevaliers which are for injuries of deed and word. The Chevaliers want it to be repaired and [to be] well satisfied in order that the enmities which could be between them may be assuaged. I will talk about some of them in order that those who would want to build agreements can take some advice on it without however wanting to teach the reader anything that he could not himself do and invent. But because I have come to these terms that speaking of quarrels and duels, and of that which can come from it that it seems good to me to say some words on satisfaction.

So when someone would have drawn the reason of an injury that one would have done him, he could say to his side that of he only he the opportunity to do this he would have been vexed to do it otherwise and that without opportunity he would have done badly and non as an honourable Chevalier and for that which has happened, he prays him not to separate from the friendship that they had together. His party would be satisfied, regretting having offended him, and that on the advise of these friends he wants to remain constant.

If two Chevaliers drawing swords one on one, of whom on stands wounded for satisfaction, he who will have wounded his companion could say to him, "I hold you for a gentleman of honour and brave valour who has made me prove it, as an honourable gentleman must do. It is the strength of arms which has thus willed it that I beg you that we remain friends." His adversary could respond to him, "Since you have held me for an honourable gentleman, [and] that you have known that I have not failed in the combat that we have had together, I am content of proving to you that I have made my courage."

If someone had offended another without reason or inappropriately, he could say to him, "I recognise my errors and admit to having done something that I should not have done, and against all the rights of an honourable Chevalier, I beg you to excuse me and want to forget it." His adversary must respond to him, "since you recognise the offence which you have done to me, and, that you repent of it, I content me of my honour and do not want of yours." If someone had insulted a Chevalier and afterwards would have fled and he who is insulted did his duty having put his hand to the sword to draw his reason, and yet could not catch him, he who made the offense could say, "I have offended you against the rights of a Chevalier of honour and valour. I recognise my fault. I beg you to pardon me for it, being assured that if I had stopped when you ran from me you would have drawn your reasons and an honourable Chevalier must do." He who is insulted must respond, "Since the you admit in the presence of the gentlemen who are here present to have fled after having hit me and that it is against the rights of a valorous Chevalier and that you recognise the act which you have done to me, and that I made my reasons for it, I want to remain constant and satisfied for it."

If someone had given a Chevalier a blow with a club, he could say to him, I have insulted you to my advantage and against the rights of Chevaliers and of honourable gentlemen being assured, if I had called you out one on one you would have done all that which a gentleman of honour and man of means must do in order not to make reason of all that which I have asked of you, I have done very badly and have greatly displeased myself. I beg you to forgive me for it. And if that which I have said to you do not content, I will do for you as it pleases you and as your friends command it." He who is offended must turn himself to these arbiters, "I have my honour in your hands. I beg you tell me if I have opportunity of being constant," which they must say to him that he has opportunity to be content. Then his party puts this to this to do him such satisfaction as it will please him. And when he could respond, "Since you know that the deed that you have done to me is against the right of an honourable Chevalier and that you have repented through the advice of my friends, I content myself wanting nothing of your honour, desiring to keep mine. If this form is not good, they can find another."

If anyone wants to make another believe that he has held words which are not true and which he wants to be repaired, he could respond to him for satisfaction that he had always felt that which he had said was always as he had heard but that he knows very well the truth and that he was very aggrieved to have it said.

If someone has spoken badly of another for satisfaction, he should say this, "I do not think of having said such a thing and, if I said it, I spoke badly and I am strongly repentant for it. I beg you to excuse me or to forget that which has passed between us", or that it is no longer remembered, ou that all that which passed between us be put under foot, forgetting it without speaking of it further nor wanting to refresh our differences. One could find many other very suitable terms on a similar quarrel and which could serve several other satisfactions for the quarrels. It could be added to all these satisfactions, both those of deed as of words, these terms. "and if that which I said to you does not content you, I will believe of it that which your friends and honourable Chevaliers will say of it. And of you would have offended me and that your would have made me a similar satisfaction, I would be contented by it.

There are satisfaction which have been created that I find very good and very relevant. Two are in a quarrel [and] the middlemen[132] try to put them into agreement and beg them to put in writing the words which they have had together, it is found that there is very little difference in their language. On this, the arbiters, seeing that they do not differ about that which they have had, advise telling them, "Messieurs, seeing the difference which is between you, and that you agree in your words, you have no opportunity to seek each other out, having both well satisfied your duty. We beg you both of remain friends as you have been before now." I think having been of the first creators in my quarters of this procedure, in these examples one can

---

[132] *les moyenneurs*

add there or make other suitable satisfaction which are in common use between Chevaliers.

Chapter XXIII – For gentlemen who had a quarrel to agree, it is necessary to know the basis and the origin of their quarrel

When one reconciles a quarrel, the arbiters must always follow the subject from where it arose. I do not speak without reason because it is a custom that one does not looks except to reconcile the evil which os apparent, which proceeds from very little, as from a simple word which has been said a great quarrel arose, the one found wounded with strong sword blows. In order to reconcile them, one only looks to the sword blows and not to the subject which forces them to enter into proofs of arms. There are some words which have been inappropriately said. These are the basis and origin of the quarrel and when such opportunities present themselves to the arbiters, and that they neglect them this does not turn out very much to their credit. It is nonetheless the true path which it is necessary to hold to in order to end a quarrel, and [I] maintain that never will it be reconciled if the arbiters do not actively look at the basis and harm of their dispute. We could very well argue the contrary and say that even if it is not needed to come to the origin of the quarrel in order to conclude it well, it could happen that the circumstances and that which could befall would be so harsh as to exceed the first subject and the of necessity he should reconcile them as much for this reason as dot he origin and basis of the quarrel. But be that as it may,[133] I will persist always that it is necessary to probe the first matter of their debate, and then after one can more easily reconcile the circumstances which have arisen from it. Otherwise, I think that they would not be well reconciled (it would be better to prevent) and when the quarrel is promptly well formed the friends advise putting them into accord in order that this quarrel will not engender greater harm.

---

[133] *Mais encore que cela fust*

Chapter XXIIII – If it is necessary to make the two gentlemen which have had quarrel embrace after they have ben reconciled

When a quarrel is reconciled, the arbiters are eager to make the parties friends and pray them to embrace as a signe of friendship. It is for this end that when they will see each other another time they may be quicker to greet each other. And in truth, this embrace is a witness to wanting to forget that which has passed between the parties or at least for wanting [them] to remain content. And when one will realise that one of the two does not want to dispose of it, one could judge that he still has resentment, or that he does not want to stand as [the other's] friend. Also, I believe that there is no great need to make them embrace because it is evident that this embrace does not proceed from a good and kindly will. For example, two Chevaliers who had a quarrel, submitted themselves to believe their friends concerning their difference. After being reconciled, one asks them if they would not embrace and stand as friends, the one responds "It is possible." He who was the most offended, "I am content to be satisfied according to the advice of my friends but, as for embracing, I will not do it, for that which has passed between us two, I will ask nothing of him nor want to be resent me for it, but I cannot be his friend." It is possible that one will not be a friend of anyone and nonetheless one will not have a quarrel with him. I know enough reconciliations which has happened in this manner. But the arbiters could say to them, "You thus are reconciled from the difference which you had. I seems to us that you have no more subjects that you are seeking. We and all this company which is here assembled pray you to remain friends" and without forcing them to respond or to embrace each other. I know well that several will find my opinion wrong. If this is so, I speak of it with reason and (according to my judgement) with truth. However, not wanting to turn away from the accustomed manner and style that serves one in reconciliations, I will conclude that in an agreement between two Chevaliers who have a quarrel, it is good according to God and Reason to put them into agreement and render them friends but not that necessity commands it, if this is not the good will of the parties.

Chapter XXV – If two kings should fight individually for the their states

This question is disputed by many expert personages who are not all of the same opinion. Some do not consent that the Kings fight individually,[134] and allege reasons to confirm their words which are good and very reasonable. The lord Muzio is of contrary opinion to all, and says absolutely that kings are bound in their own person to fight with the sword and one on one for their Kingdom against he who wants to seize it, without the subjects participating in this quarrel. I cannot pass this opinion in silence. I believe that the Estates of the Kingdom will never allow that a king hazard thus his crown and his country. Several reasons are apparent why a King must not fight another King one-on-one because this would risk his Kingdom, his person, his children and his subjects. If he came to be conquered, this would be a notable house[135] extinguished and his Kingdom lost because his subjects do not want to take up arms for his defence. I believe also that all the states of the Kingdom would never consent to it, even a King who came legitimately who succeeded to the crown. But I think that Muzio favours the fight of the line of Amadis of Gaul,[136] who by their individual fights, Chevalier on Chevalier made proofs of their prowess. In those times it was superb, if one wants to add faith to it, that on regards by all the discourse of monarchies one will find that they have been debated and conquered by force of arms. Nynus was King of the Assyrians and his successors enjoyed it for a long time and through discord which was between them Cyrus was rendered King and Monarch of the Assyrians and Persians, after which his successors enjoyed it for a long time. And through the discord between them Alexander the Great, after having subjected the said Persians and Assyrians conquered the flourishing monarchy of the Greeks, and after his death all these Kingdoms were pounced on by many of their lords. Then the Romans made themselves masters of all, that which remains stood with Caesar and his successor Octavius, which is the fourth monarchy. All these countries of which I have spoken, these Kingdoms and monarchy have been conquered by force of arms, and not man on man and king on King. Also, I believe that this may be

---

[134] *combattent particulierement*
[135] *vne notable race*
[136] Almost certainly the *Amadís de Gaula* by Garci Rodríguez de Montalvo, an important collection of chivalric romances popular in the sixteen century but written some time in the fourteenth.

seen as great imprudence to put a kingdom at risk[137] for want of know how to defend it, a cowardice too great that they subjects did not oppose it with lively voices. King Charles the Seventh, King of France, found himself relieved of the greatest part of his Kingdom by the English but with force of arms he took away all that which they possessed and they have not returned since, being well attended by the lords of his Kingdom and by his subjects. We ask God for Kings to govern us and who have commended us to obey them. Since we have a King, it is necessary to honour him and obey all his commands and aid him to preserve his State. Otherwise God will send misfortune to the subjects who have not obeyed and defended their King.

---

[137] *en proye*

Third Part

I have spoken amply enough about the callings out which are [made] between chevaliers and have deduced that which seems to me to be proper for this subject. I have also brought out that the chevalier must not fear to ask for pardon for an injury to whom he may have done wrong and against his duty, together with the satisfactions which must be made in both word and deed, and whether two Kings must fight one-on-one for their Kingdom. We will speak at this time in this third part of the arbiters and their qualities, and those who must be called to decide quarrels, with other questions and requests which are proper and useful for this subject of chivalry and worthy of an honourable chevalier being instructed in it.

Chapter One - Of the status of the arbiters and their quality

It is very difficult to find arbiters who are appropriate and discreet in order to conclude and end the agreement of the quarrels which they follow. Yet all feel themselves capable of exercising this honourable vocation. So in order to speak about it, it is necessary to choose an arbiter who is an honourable gentleman, wise, discreet and very learned. It is a very praiseworthy thing to find an arbiter in this condition. Also, one judges this as being a gift of nature when people are accompanied by such perfection. This is not to blame anyone. I am assured that one will confess to me that there are those who are full of good counsel and more ready to give good advice. As one sees with the better jurisprudence[138] and more certain opinion to which many honest men yield and prefer them to others, so it is with the great captains that the Kings choose to lead their armies, who are experienced in busi-

---

[138] *iurisconsultes*

ness and handling and conduct of arms. This comes as a gift from Nature, who wants to distribute these faculties to such people. This is why it is rare to find many of them. Also, when there is a quarrel in a country between two gentlemen, it is necessary that the parties take to the wisest and experienced in deeds of arms to better know how to debate the fact of their part. This is why I am of the opinion that it is difficult to dispute the right of his side, even when it is a question of a quarrel where there is dishonour that the one who is an arbiter is not well versed in war,[139] or that he has not seen and heard the practice of reconciling and end quarrels well and, as they must, because it is the place where one hears honour spoken of. One has to believe that when a gentleman has declared it, he is much more distinguished and honoured and able to talk about reconciliation than one who would have rested in the private affairs of his house. And even if one took an arbiter who was or had been quarrelsome, he should not for that be rejected from the company of arbiters. There are those (according to my judgement) who better disentangle the business and better pick through the circumstances and recompense better the points of the quarrel because it has happened to them and they are experienced in it. But I hear that they would have divested their first actions and that they would have changed them into a cheerful and softer humour, as there are some who have seen very great quarrels who have made themselves honest men, more temperate and of a modest humour. And if it is possible, take an arbiter who is a gentleman. I should very much fear that those who take others are wrong, not that there are no honest men who have the title of captains, and who are wise and discreet and of good advice. But the quality and the authority of the one who is a gentleman very often lessens the quality of the one who is not and, coming to disdain him, this is the occasion that the accords and agreements are broken or are not so well made or examined. There are also made reconciliation that the parties want and consent to have occur in front of a great lord, or a third party who serves as a superior to judge it. This is very expedient in certain quarrels which may be so difficult to deal with to parties so opinionated that it is necessary to have a lord[140] to make a judgement and to determine it with the opinion of the arbitrators and the parties. But also it would be necessary that he who is thus delegated to be a third, does not favour the one party more than the other, and embrace this affair so skilfully in order to judge it according to duty and reason, that he makes them content. And if he did otherwise, he would not be wor-

---

[139] *n'ait hanté la guerre*
[140] *vn grand*

thy of being called to such arbitration nor to be elected to hold the rank of superior judge, and in doing so would be judged unworthy.

Chapter II - That it is required to know the name of the arbiters

This is a custom which is very bad that those who have a quarrel, when one speaks to them of reconciling and of going to arbiters, they hold the secret wanting that they may only be known on the day of the arbitration. I am of the contrary opinion that it is necessary that the parties name their arbiters and that they be known. Otherwise, it could happen that they may not be of similar quality or they may be enemies, doing this would make two quarrels of it. But it is necessary that the arbiters be friends or at least that they know each other. I will be therefore of this opinion that two who have a quarrel, before they find themselves together on the day of the arbitration, that the one and the other choose arbiters who are equal, if it is possible, in honour, rank and house and friends. Yet one must not doubt this honest frequenting and knowledge that the arbiters have one for the other.

Chapter III - Of those who effect a reconciliation for an accidental quarrel[141]

There are gentlemen of such honest condition that when they know of a quarrel between their relatives and neighbours, they act promptly and unexpectedly, and they instantly present themselves to take the words of the two quarrellers and, after having taken their words, pray them to not seek each other out unless they have taken a day to reconcile through the advice of their friends. This procedure is very commendable and worthy of a gentleman of honour. I would want that this honest means of doing [things] was continued. One would not see the nobility so often call themselves out to combat as one does today. It is necessary that the arbiter conduct himself there so well and with such skill that, taking the word of the two contenders, they are neither surprised nor offended, because if this happened it would not be to proceed as an honourable and valorous chevalier.

---

[141] *vne querelle suruenuë*

If in a quarrel which arose between neighbours, a gentleman, a common friend, in order to prevent them calling each other out takes the word of one and postpones the other until the next day. However, this latter will make a challenge to his enemy who had given his word first. I ask as this challenge must be called. I will say that the arbiter has done wrong by the first who gave him his word and that he is not obliged and must revoke it as being deceived by it and can dispute their quarrel, as should have been done previously. And no matter how much the former has given his word, I maintain that he is well called, and that he must satisfy it promptly and at the same time, or else he would be wrong in his honour, and the mediator to whom he had given his word is not offended by it. Although the one who had given his word might complain of him, because there would be an appearance that he embraced one party more than the other, which a mediator must be careful not to do, for there would be in it not much honour and less reputation. It is necessary in such affairs to be faithful and not passionate, or otherwise the arbiter should be named a deceiver. I will bestow in this place advice to those who would want to reconcile an agreement for a quarrel. It is that he who must make the challenge must be the first to give his word because it is he who is the first offended and it is necessary that the arbiter take his word first and that he holds it for certain and well assured, and then after he will take the word of the other. It is not reasonable to take first the word of he who has done the offence. Then he has nothing to ask him. But he should take the word of he who is offended in order that he who has done the offence may not be surprised by it and after that both will have given their word that they proceed chivalrously and not dispense with the honour of chivalry or from their promise, and if they would have done it otherwise, the arbiter must complain of this wrong that they do to him. The word between chevaliers must be true and a chevalier must not fail in his work to another chevalier, if he does not want to be held as perfidious and villainous.

Chapter IIII - That one must keep his promise between gentlemen

There is nothing so praiseworthy nor which must be as much esteemed than the promise, which must be so dear that when one has made a promise it should be held inviolable. It is the office of an honest chevalier. However, if the promise is not reasonable, and it contravenes the duty of a chevalier or it is not within his power to be able to keep it, it is excusable. As the promise is not being a forced thing but free and voluntary which does not exceed beyond the ability because no one is obligated by an impossible thing. Also, on the contrary, if you have promised your friend not to hold the side of your enemy and nevertheless you accompany him [ie: the enemy] and make him happy, thus your friend, to whom you have made such beautiful promises, has reason to complain of you for you have done him a perfidious deed. Therefore, promises between gentlemen and friends must be kept. There is nothing which obligates a chevalier as much as their word. When it is said and promised, it must be kept. I say, moreover, that it must be held by anyone. Augustus Caesar had published that whoever could take Crocotas, the greatest and most experienced thief who ever was in all Spain, to him there would be given twenty five thousand ducats. Crocotas, being made aware of this, presented himself to Augustus and asked him for the twenty five thousand ducats. Caesar, in order to show that he must keep his promise without having regard to the merit of this thief, had them given to him – and also [gave him] his grace. Messius, who was dictator of Albania, because he had not kept the promise which he made to the Romans, was drawn with four horses. Thus, as the kings who made promises are condemned and they condemn themselves with very strong reasons, that he who carries the title of gentleman must have a chaste and true word. His promise must be inviolable and, if the chevalier is false, he makes a great stain on his honour which is irreparable. Let the chevalier take, therefore, care not to engage in so many words that he cannot uphold his promise.

Chapter V - When a gentleman has given his word, whether he is bound to name he to whom he gave it

This request is not without difficulty. Many are of the opinion that the gentleman who has given his word, which is of important consequence, is required for his justification to name the one to whom he gave it, to prove that he did not contradict him. There are others who opine that he must not be thus constrained. I would be of the opinion that when a chevalier has given word, he cannot be in any way obliged to name he to whom he said it, if he it does not please him. It suffices that he says, "I am a gentleman, a strong man of means and true in my words. I invent nothing nor would I want to do that which could defame the good reputation of another. That which I tell you holds truth." It is severe to force an honourable chevalier to make him say such words if it does not please him. He should not, however, conceal that which matters to his friend without alleging from whom he was told it when, to me, I approve greatly those who warn their friends of that which one says or one does against them, and those do not do the duty for true friends who do not warn them of that which is done to their prejudice. There is no man which can keep me from warning my friend of the harm which one causes him nor of that which one says about him nor which must force me to name him of whom I hold it.

Chapter VI - That the assemblies must be defended by all the Provinces

The King must be desirous of tranquillity, harmony and union for his people. He must also ensure that the quarrels which arise between the subjects are appeased and prevent that they go so far such that one may be constrained to take up arms. To avoid this occurrence, he must forbid all illicit assemblies and give an express commandment to all his governors of countries and provinces to allow that none in their government suffer it. That brings back an infinity of mischief and murder. Because, if there is a quarrel in a country between two having authority and friends, each one uses all the means he can in order to defend it in such a way that there is no one of quality who can prevent him from this. He proceeds from it to other quarrels, so that in a government you will see more than twenty [quarrels] having emerged from one. And if there was an authoritative governor who was stayed on the spot to suppress this sudden fury and expressly order everyone to lay down their arms and stand in their house without making any assembly touching such debates that he does not have the recognition to reconcile them, this would be very beneficial for the public. In order to do this, it should be necessary that the King makes a solemn edict, through which he forbids all illicit assemblies for whatever reason and occasion that may be and that he wants and intends that his governors take cognisance of all the quarrels which arise in the jurisdiction, and that no one be so daring nor so brave as to take up arms on pain of his life. Also, the governor of the province must know if there is no quarrel in the country and if he is warned that there may be some, he commands the parties to come before him in order to understand their difference and reconcile them. And if their quarrel is of such consequence that they are forced to come to arms, or if they do not want to obey his commandment, then he should rigorously[142] order that if they do not put down their arms and if they do not forbear from making such an assembly that he will run them according to the King's will. However, it would be very wisely done to warn the King of this tumult so that it pleases His Majesty to administer it when the King demands the two parties to come to him. Then, if they do not want to cease nor to obey his orders nor the governor of the country, [he] corrects them very severely and sends them back in agreement to their houses. And if they do not want to obey the King's order, that is a capital [offence]. If it were done in this way, you would see in the Kingdom everyone at peace and without quarrel.

Chapter VII - What forms should the governors of provinces follow in order to appease the quarrels arising in their government?

I shall speak at this time of the form and the manner in which the governors of the provinces must follow to resolve quarrels, if his Majesty would agreeable to it. It is that when the governor wishes to have a quarrel resolved that he would bring to light to the parties and in the principal city of the country, without the parties being required to have an arbitrator for them. But it would be necessary, in place of arbitrators, that the governor should come with him six of the most renown and notable and experienced gentlemen of his government, who he will have commanded to decide quarrels, and that the parties would pass before their adjudication. And these gentlemen, who accompany the governor, should be appointed by the King, to whom his Majesty would give wages every year, which would be paid to the appointed receivers in receipt by the said government. It seems to me that the King could not make a more worthy and honourable expense than this, and which would be more for the benefit of the fatherland. This assembly could be called the chamber of arbitrators, or the council of the province, which would be held four times a year or more depending on the circumstances, in the principal city of the government where the governor or his Lieutenant would preside, whose parties would be held to believe and to follow the advice

---

[142] *rigoureusement* - Cotgrave has "Rigorously, sternly, austerely, severely, in extremity, without any favour"

which is given. And if the quarrel was of such great consequence that the parties did not want to consent to the statement of the governor and his council, the governor must send them back before his Majesty with that which has been deliberated in order that he knows that one has definitely proceeded in their agreement, otherwise it will be received and amended by his council, if necessary, for whatever reasons necessary.

Chapter VIII - That an honourable chevalier hearing [someone] speaking badly of his friend must respond to it

Scipio Africanus said that there was nothing so difficult as to keep a friendship forever and until the last day of one's life. But Julius Caesar, as Suetonius recounts, kept the friendship he had for someone constantly, and only lost it with great difficulty. The friendship that we promise to bear to each other must be so loyal and so faithful that it must remain perpetual so that in our business of necessity we can be assured of the fidelity of a good and legitimate friend to help, advise and favour us when we will necessarily have to deal with that business, both in our absence as well as in our presence. And this friendship should not be ended for small and light occasions. A good, faithful friend, round and whole, is worth as much as a rich treasure, and one cannot desire greater riches in this world. The good friend who will be faithful to you will always try to show you testimony of the entire friendship he has vowed to you in such a way that it is very reasonable that this friendship should be reciprocal and mutual. It turns out that the friendship of Amon and Phisias was so great that they, two noble captains, [when] one of them was taken prisoner by King Denis of Sicily[143] who sent him to Syracuse, wanting to put him to death. But he begged the King to allow him to go to his house to put his affairs in order before dying, and that he would warn his companion to pledge of his faith, which the King promised. When the day on which he was to arrive had come, and he did not come as promptly as he had promised, everyone laughed at the one who had guaranteed his companion to have thus committed his life. But he replied that he made sure of his companion's friendship, and that he would not fail to return, which he did at the same time as he had promised.

---

[143]Dionysius I, the Elder, c. 432 – 367 BC, tyrant of the Greek colony of Syracuse was regarded as the worst kind of cruel and vindictive despot.

The King, amazed by such great friendship, pardoned the offense of him who was a prisoner and begged them to receive him as the third in their friendship and as [their] companion. This, then, is how this friendship, thus fortified by many proofs, must be tested even more by the absence of his friend. However, the greatest demonstration of friendship that the friend can make for his friend is when he finds in some companion, who speaks badly of him, and takes up his cause and maintains the right of his friend with affection. It is the office of a good friend to answer for his friend in his absence. Also friends are known principally in adverse things and are of great importance, because everything we do for our friend results from a constant friendship and the good view that we have for him. Tarquin the Proud, when he was driven out and exiled from Rome, said aloud, "At this hour, I experience how many good and faithful friends I have, and also how many unfaithful ones I have, and to the one and to the other, I am misheld by them, my friends. At this hour, they flee from me when I want to search for them." The Poet says, "When you are happy and all your affairs are well, you will find many friends. But if your good fortune changes into bad, you will be alone and abandoned by everyone." Varro, speaking of friendship, learns of testing a friend and says: "Do you want to test him? Pretend to be poor and calamitous. You will thereby know those who love you and who will be quick to help you." Also, Cicero, in the book of *On Friendship*, says that "there is nothing so proper to nature nor more conducive to prosperous and happy things than friendship. There is nothing so agreeable as having a friend with whom you could surely confer as you could with yourself." The friends also of whom I speak and who respond in this way in the absence of their friends are rare, and not many are found who are inclined to perform such honest offices. I would therefore conclude that any man who hears [someone] speak badly or who knows that one brews something against his friend, who is against his ruin or to make him displeased, he must warn the friend and respond in his absence and work for him.

Chapter IX - If a prisoner of war, having given his oath, must keep it

It is a maxim that when a prisoner is taken in a good and legitimate war and, having given his oath, he must keep it. And falsifying it, one can call him a perjurer. There are, however, many reasons which are apparent and contradict this law. It is therefore necessary to examine in what condition the prisoner gave his oath. I will say first of all that the prisoner being put to ransom, knowing that it does not exceed his strength and power, is bound to keep the oath he has sworn. If he did otherwise, he would be condemned by the Prince to present himself and to pay the ransom he has promised, especially since this compact is made by mutual agreement. And if it were that the prisoner had been forced to give a large ransom which could not nor would not be in his power to provide, whatever honest remonstrance which he may make notwithstanding, that he is forced to give a large ransom and to give his oath, my opinion would be that he is not bound by it, because it is a constrained and forced promise that he made to recover his freedom, which must be held as nothing, and absolved from the oath he promised, because one has not treated him like a man of war. In ancient times, the opposite was held. Even the Romans, who were very experienced and well advised in the administration of their republic, were otherwise, as is testified by the example of the Consul Postumius,[144] who, leading an army, was taken with many captains who were all dismissed under their oath. This was disputed before the Senate to understand if they were obliged to do so, especially since it was an agreement that was made in war. It was answered that Postumius and all the prisoners could not negotiate any condition of peace without the express consent of the Senate and of the people, and their oath was only spoken by force and compulsion. But, on the contrary, they returned to the enemies, those who had sworn peace, to dispose of their lives which the enemies returned. Also, Postumius remonstrated with the Senate that the treaty which had been made with him and his enemies was only a simple promise which obliged him, if not those who had [also] consented. King Francis I had practised this law better and more closely, which in the Treaty of Madrid he had promised to Emperor Charles V many articles which were not reasonable, which was the occasion that the King, being out of prison, said to the ambassadors of the Emperor that he was not bound by the promise which he had made, especially since the conditions were iniquitous and also that, not trusting his promise, he [Charles] had taken his [Francis'] children hostage. I will say, therefore, that all prisoners of war, kept and confined, whatever promise he has made, he can escape [it?] without being blamed but, being at liberty under oath, he should only do so when the conditions are reasonable. King Pyrrhus, who had won a battle against the Romans and taken great quantities of prisoners, gave them leave on their oath on the condition that they re-

---

[144] Spurius Postumius Albinus Caudinus, fl. 4th century BC?

turn. The Senate published on pain of their lives that all prisoners had to return on the ordained day, but not one was given [as] hostage. Any man, therefore, who has given his oath must hold it, whether Prince, gentleman or other, whatever he may be. Also, the Prince, giving his oath to his subject, must hold it and strictly keep it. I will bring back an example of the Lacedaemonian King of Sparta, who went to find the King of Persia in his Kingdom. There was a great Persian lord who took up arms against his King. He brokered a reconciliation but, since the King of Persia having his vassal in his power wanted to put him to death, the Lacedaemonian prevented him from doing so with vile remonstrance not to enact his cruelty, and that he would do harm to his greatness, at this hour that the Persian lord had declared himself his servant. I will bring back another example worthy of memory. Sultan Sulieman, Emperor of Constantinople, sent one of his Bashas to Italy where he landed at the port of Castro. The astonished inhabitants went to him that on his oath their lives would be spared. However, he was so disloyal that he put a part of them to death and the rest he took with him. Sulieman, learning of this wicked act and of his perfidy, had him strangled and sent the prisoners back to their country. A very worthy example to a Christian of not breaking his oath as even a barbarian knows to keep it. Even an oath given to thieves and pirates must be kept. Pompey, making a peace treaty with the pirates and corsairs, gave them the oath and security that the Senate had in order to be agreeable and ratified it to maintain the honour of the Romans. And if they had not kept the oath that Pompey promised them, they would have ransacked and destroyed their honour. All these examples will serve me to show how an oath must be kept, even in times of war, and under what conditions the prisoner must keep it.

Chapter X - Of him who leaves home to fight a battle, and cannot be found on the appointed day

The valiant gentleman, when he presents himself for battle or some good effect of combat, prepares himself diligently in order to be found there, and believes that there is nothing so glorious as to be found in so honourable a place. Also, certainly, it is the most gracious and dignified demonstration that the chevalier of honour can look for. It happens, however, that when one has travelled to be found in a duel that the other did not appear on the given day and, however much this was true, some will want to say that they were there. They ask how this should be avoided. I will say to make it clear that he who is prepared and has travelled to appear in a duel, event if he does not arrive there on the day of the fight, he must not be excluded from the rank of combatants, and that he should be placed in the ranks of the most valiant, as a noble chevalier who seeks the honour that all chevaliers have always sought, in order to acquire a good reputation and for his posterity. This has been observed by all valiant men, because the will should be as much esteemed as if the effect had followed. The fairly fresh examples of two battles which have been given in this Kingdom bear witness to this, the one of Serizolle where M. d'Anguian acted as Lieutenant General, where several lords and gentlemen, ran to be there. Some arrived there the day before the battle was fought, some others on the day itself, yet others the next day with extreme regret at not being there on the day the battle was fought. King Francis the First held them in as good esteem as the others because (so says this Prince, full of greatness and generosity) that they "have shown me the affection they have in my service, that they are with M. d'Anguian in order to do me further service, so I want them to be among the number of Chevaliers of Honour like those whom I hold in high esteem. Look at what difference there is between those who have not left their house and those who go to be found there." The other was that of Dreux, where M. de Guise fought all day. The day after the battle, or two days later, seven companions of gendarmes arrived there, where those of M. Latrimouille, whom I was leading as his Lieutenant, were of this number, wanting to demonstrate the regret I have for having been found on that day of the battle, together with all the others. He told us that we had done as much service to the King as if we had been in battle, and that he would report to the King our diligence, and that we had come enough in time to do something good, because the enemy is not far off, he says, pretending to come to battle again. This is how those who are looking for a fight and who are going to present themselves on the day of a battle must do so as much to acquire honour, and must be as much esteemed as those who are there. This is to clear up a quarrel, if it had arisen on this occasion, and not to object to a gentleman of honour not being found there, when he has put himself to his duty and is in the army. Those who do such offices speak more out of envy than they do with fine reasons. Certainly the noble and magnanimous heart desires its honour and a good reputation. This is why the man driven by praise asks for no other recompense of virtue than the reputation of being well esteemed. Cicero says that valiant men, and those who are prudent and wise, do not work so much as much for wanting to exercise virtue in wanting to receive a great reward, than for the honour they hope [to gain] from it. Also, at the battle of Jarnac, when the two armies confronted each other, there was only the guard who fought and the battle, which was with Monsieur, the King's brother, who has since became the third King Henry, who did not fight nor the rest with his cavalry. I was there with the company of M. de Latrimouille. I can speak with the truth about it. It would therefore be necessary to conclude that those who were in battle with Monsieur, the King's brother, were not in combat. They would be too unreasonable to want to remove the honour [from] too many good and honourable people.

Chapter XI - In the conduct of an army, which is most necessary: the bold man or the wise man

These two virtues are some praiseworthy in war and of such great value that he who is well versed in them is held for a very great captain. And if there are found captains who have both, that is rare, because it is necessary to speak about which of the two is the most necessary and judge it according to whether the art of war merits it. I will always take the advice and good counsel that a wise man is well required for the conduct of an army, and on him one must put many matters. But for the execution [of it], boldness is much to be prized. I mean to speak of the bold, strong man of experience, and also of the virtuous boldness compounded from the two extremities of cowardice and temerity, for these two are the extremities which are villainous, and from these two is composed the virtue of boldness. The bold man, therefore, who has experience and who knows how to command himself in combat without risking himself, who with brave judgement knows how to choose his advantage, going valiantly to the charge and making fight all these well-ordered troops, must appoint a great captain, having waited well for his advantage, to fight his enemy, and himself fought valiantly. He should be highly recommended to lead an army, and much more than a wise man who does not have the strength and perhaps the heart to go into battle. And although he is wise and a man of great judgement, if he does not have this virtue of boldness, he cannot be considered a great captain. I have memory of having heard it said to Monsieur de Guise, Francis de Lorraine, that the first art of war was to be bold, believing that a valiant man, well experienced, will obtain victory more often and execute it better than the wise man who has no boldness. Monsieur de Guise, of whom I have just spoken, gave enough proof at the battle of Dreux of his boldness and his experience, and if he had disbanded, he who was leading the guard, when the battle was broken, and the cavalry put down the road, the battle was lost. But he stood firm and did not leave his rank and fought all day with his arquebusiers. His experience and his boldness were the cause of obtaining the victory. Marshal de Tauane, was he not praised for his boldness, for his long experience, was called near the person of King Henry III to the conduct of his armies, where he fought two battles and had victory. I will speak of the Marshall de Brissac, who was praised and highly esteemed, who had made fine exploits of war in Piedmont up to the gates of Milan, but without experience this boldness often turns into temerity and is a good cause of losing battles. The lord of Nemours, leaving the house of Foix, experienced enough how much great boldness was harmful to him, who after having won the battle of Ravenna, wanting to pursue victory with three hundred horses, lost his life pursuing a squadron of Spaniards who were joined together to make a fine retreat. What I am saying is not to reject the wise captain and good adviser in the conduct of armies, but my opinion is that the greatest strength of an army is to have a bold and experienced leader. The ancients held a chief of arms in great esteem when he was accompanied by valour, good experience and knowledge. Cymon, the great captain, used to say that there would be better an army of serfs led by a lion, than an army of lions led by a serf. The general of an army should be learned, or at least of a beautiful and brave speech, to know how to admonish the soldiers of their duty. Caesar was the very expert there. It is held that these harangues which he made to his soldiers in full armies have indeed caused him several victories. Agamemnon highly prized and highly praised Nestor's advice. But also if he had not been accompanied by the boldness of Achilles, Ulysses, Ajax and several other bold captains, Nestor's advice would have remained small. Alexander had Parmenion and Antipater and other great captains for his advice. But his boldness and his good leading with enthusiasm was indeed the main cause by which he obtained so many beautiful victories. Such is the boldness and the main force in the war.

After having spoken well of this beautiful virtue of boldness, finding myself on the subject of speaking about it, I must put forward what I have seen practised. It is that I have known enough bold men to draw well the sword who were not resolute and determined in fighting the war. Others fought well and resolutely, and who nevertheless would not have wanted to beat each other with sword blows. I can testify to this and have never been able to find the reason for it. If it was only the great companion when one goes to a charge, it is the occasion that those who do not want to do it with sword strike go there more determinedly, the others who do it with sword strikes fear the charge, because they do not want to do it, are not accustomed to it. But I will always say that any man who draws his sword well and resolutely is considered a bold man, and must not be doubted that, going to war, he will not face it when they would experience it and get used to seeing it there. There are some who are born so bold that from their youth they go into battle without apprehending the peril and the evil that can happen to them in the long run. They recognize themselves and correct themselves for their temerity.

Chapter XII - On the difference that must be made between gentlemen who claim to be of better house than another

There are great quarrels over this subject, and there are not many gentlemen who want to endure being told that one is from a better house than them. If it is to be believed that all gentlemen are not alike, neither in quality, nor in house, nor in fellowship, it is necessary to yield to the greater. I wish to clarify and examine this point about a good house, which is not without difficulty and which must be well understood. The gentleman therefore who is of an old house has something more than another could have, being that his old house, without having changed its name and arms, and possessing large and beautiful seigneuries, and that from his house there have come great personages who have had honourable offices, have led armies and won battles, and are ordinarily called to the service of Kings. Truly, to those, other gentlemen must yield to them to recognize them as gentlemen of honour, of respect and coming from an ancient house, as appears in their genealogy of five, six, seven or eight hundred years. In these great houses, the Princes do not disdain to enter into fellowship. There are also enough gentlemen who are not so great in wealth who nevertheless do not want to yield to someone greater than themselves being of ancient lineage, and well akin in nobility, name and arms. That is the reason why a gentleman of honour and of an old house would be wrong to endure anything that offends his honour, from a greater lord than him, that is why when I speak of a good house, I do not intend to speak only of goods, but I mean to speak of a good and old house. It is true that when it is accompanied by many goods and beautiful lordships, it is even more illustrious. However, it could be that a gentleman may have come to the succession of a good and large house, accompanied by beautiful, large and honourable estates who previously had very few goods. Life will recite in this place an example of two Gentlemen of good house: one was the Constable of Clysson, and the other the Lord of Cran. They had a quarrel over some nasty remarks they had made to Monsieur d'Orleans, brother of the King, which quarrel was decided in Paris. We know well how the outrage made to the lord of Palaizeau by a lord of the house of Rouen ended. This is how gentlemen of good house do not want to endure an insult from one greater than themselves, but also, failing this, gentlemen must honour each other and according to their quality. One must not esteem the other more than one should. He who has acquired honour and reputation cannot suffer to be accompanied by someone less than him and who has no merit. There is much more. Not all gentlemen can speak of being from a good house. There is some distinction which is apparent. For of one it will be said that he is a great lord, and that his house has always been old and greatly allied. The other one will say that he is a gentleman of good and old house, in name and arms. Another that he is a gentleman of honour and of good breeding. Yet another that he is a gentleman and a good man. By this, you see that gentlemen are different from each other and cannot be equated and yet if one of these is offended and insulted by someone greater or lesser than himself, he want it to be repaired. I am of this opinion that the great and oldest houses of today in France have been made and elevated by Kings, and by the services that they done for them and from lineage to lineage have maintained themselves there with honour, and also by the great alliances they have made in their houses. There are, however, many great houses which have changed lineage and those who possess them today or take the names and arms of this house, have made themselves great lords. There are others who have not long been gentlemen but are nevertheless rich having fine seigneuries, but they have not held these beautiful houses or riches from the antiquity of their houses. I nevertheless want to give a good reason for what I have said above, that it is not proper nor very honest for a gentleman to say to another that he is of a better house than him, and the honest, well-trained gentleman will never speak in those terms, except if he is greatly forced to do this by someone who wants to make a difference on him and to make some comparison or to equal himself, and that he knows obviously is not his equal. There are also some who are so promptly elevated and have made themselves so great that they want to be honoured and respected. These honours must be reserved for the greatest. It is a fine prudence to know how to know oneself and not be desirous of an honour that one well deserved and leave it to those to whom it belongs to do so.

Chapter XIII - Continuation of this chapter and where this name of gentleman came from

We hold in France that he who is called a gentleman must be a gentleman by name and arms, as I have said above, and can only possess this name by the ancientness of his house, and not by loan or newly having acquired it. This was proven by Cicero in calling "gentle" those who are of the same name, and who have always been of free condition, so that no one of their lineage was ever a serf or a slave, nor degraded in honour. Boethius also says that, in ancient times, one called "gentle" all those who were members of an ancient house and lineage, such as Brutus and the Scipios, and the other noble houses of Rome. Therefore, this title of gentility was attributed only to noble houses and to their gentlemen of name and arms. There are some in France who have made themselves great lords, like Counts, Marquis, Dukes, and Peers of France, and, of those, there are some who are not so great and yet are gentlemen in name and arms and ancient house but luck and prosperity did not accompany them so well as others, so much so that you see several houses that were once great and rich and are now ruined and destroyed and others, who were small, today greatly elevated. It is the change that is taking place in this world which we are all bound to obey and to yield to. It is the wheel of this century which has thus determined it. We cannot resist it. Nobility was acquired formerly by virtue, and those who were ennobled by arms were the most prized. Also, the Romans gave these ones great privileges and allowed them to wear signs for their coat of arms in recompense of the victories they had obtained. We still see today that gentlemen carry in their heraldry a shield and several want that one holds them in the rank of chevaliers. Today, all the nobility wants to be similar, and want to be given as much honour as if they had conquered a Kingdom. His Majesty should put order to such sumptuousness, and among other things, only allow this title of Lady to be given to women whose husbands have acquired it, and been given it by special grace and liberality of the Prince.

    I have seen in the great houses of this Kingdom that this quality was not given without the will and permission of the King. This was done for the great services he customarily drew from his nobility and, wanting to reward them, honoured them and their family by making them greater above the others, and more likely to serve him. And if the King does not give them his order of Chivalry, he generally made them all in full army, holding them for chevaliers, and then their wives carry the quality of Lady. At Lendrecy, King Francis I made all the gentlemen of his cornette[145] of his household. King Henry II did the same at Rency. This dignity was called "Chevalier of the Accolade." But to the captains of his *gendarmes*, when they had done him great and signal singular services, who merited being recompensed, and to ambassadors who had done him long service and who had been employed in great and beautiful affairs, to these he gave the regalia[146] of his order. In short, this title of chevalier is today of such credit that only the greatest are held honoured to be called chevaliers. I have seen in the time of King Francis I and of King Henry II, a captain of fifty men-at-arms, even though he was not accompanied by much wealth and great lands, being respected and honoured by a lord of fifty thousands livres in rents, even Princes, so much honour and virtue were at that time in the kingdom and in singular recommendation. This is why the antiquity and the generosity of a house are much to be esteemed, which shines only in certain species of men whom God and nature have chosen from among others, who cannot be common among men, especially since it is a rare thing to find so many illustrious and noble persons. And although all gentlemen tend to this point, they cannot perfectly reach it. It is reserved under a certain providence to some in particular to excel above others.

---

[145] Dictionary of the *Académie Française*: Cornette: is also a long and wide strip of taffeta that the Councillors of Parliament formerly wore as a mark of honour, and that Francis I granted to the Professors of the Royal College of Paris.
[146] *collier*

Chapter XIIII - Whether someone is doing himself wrong if, being outraged, he employs someone greater than himself.

It is quite common that he who has a quarrel with someone greater than himself takes the favour of some great lord to assist him in order to be supported in all his undertakings and to make his part stronger. I cannot praise or value this greatly. He who undertakes to avenge his friend's quarrel, or to assist him, I believe that he is doing himself great harm, and still he is doing it more if he takes any other pretext than the quarrel of his friend. If he does it and it is with advantage, he acquires a dishonour and a perpetual quarrel, from which only much misfortune can result. I do not want to say that he who employs a greater person to assist him in his quarrel does wrong, because in a quarrel the weakest must seek support and favour, which authorises him further. Also, he who is thus employed must be so advised not to undertake a deception for whatever prayer his friend may make to him or obligation there may be between them. It is up to Kings and Princes to maintain their servitors and to avenge them for any injury done to them. Yet it would be better to try to get them to agree than to embitter them even more. The Kings still favour the party of someone, and they themselves want to avenge the quarrel of the one they love. Very often they suffer for it. Philippe de Commine tells that Henry VI, King of England, for having supported the house of Lancaster against that of York, lost his status and was killed by his victims. The Marquis de Peghaire took no other circumstance to conspire against Emperor Charles V other than he kept the Viceroy of Naples against him, all the more so because the gentleman, however great he may be, takes good care when espousing the quarrel of another.

Chapter XV - If he who is envious of another must be held as an enemy

Plutarch is of this opinion that it is very difficult to avoid the envy of others. Thucydides esteems envy to be the necessary companion of authority and power, and says that one is led by good counsel, which in grave and important matters chooses that which is most sudden and suitable to the urge. We discover enough that there have always been envious people of whatever status and quality, and especially between neighbouring gentlemen, and even parents are most often jealous, making good food and good pretence, behind their faces, actions that sometimes turn out to be consequences. He who is envious of his companion always brews his ruin and prevents him if possible all his plans and his advancement. In the end these are only dissimulations, that is how envy is very dangerous, and must not be proud at all to those who carry you away, but they must be deemed as private enemies. There are many cravings that are taken in good part that the Latins call *emulatio*. Those are driven by an honest jealousy to see their companion doing well and are eager to follow them, take care in their gesture, in speaking well to them, in their grace and honest bearing. These must be prized by each one having affection to follow these honest people in their perfections, proposing to imitate them and to acquire as much praise. This desire is very honest, because properly *emulatio* is to follow and imitate the perfections of others and to try to acquire as much honour and praise: if each one were desirous of resembling those who are well furnished with them, we would then see many gentlemen of honour flourish, and it would be better for those who could do it. This envy that I speak of here is very commendable, and I advise all honest gentlemen to follow it and forget that vicious envy, which is the total ruin of honest men, an envy full of malice which has no friendship and only loves itself. I will recite a quarrel that ended at the Court of King Henry II between the Baron des Guerres and the Lord of Luffebourg[147]. Their quarrel arose on the occasion that Luffebourg aspired to be master of Monsieur de Lorenne's wardrobe, as the Baron des Guerres also did, and knowing that the Baron was preventing him in this, begged him not to oppose the good that Monsieur de Lorenne wanted to do for him. The Baron excuses himself and tells him quite flatly that he is no power in this. Luffebourg, well angered, draws his sword and kills him, telling him, "Here is the recompense for having chased me from my master's service," and fled. We see in the chronicle of Charlemagne, Gauves, for the envy he bore Roland, was the occasion of his defeat and his death. Envy proceeds from ambition, and ambition is nothing other than being envious about the honour, advancement and glory of others, by which we can judge that there is nothing that dissolves a friendship so much as envying him. From this a great enmity is engendered. We can see enough from the stories of how much harm envy has brought to monarchies and republics and to all people who wanted to be biased by ambition. During the reign of King Charles VI, the ambition

---

[147] Offenburg?

of the Dukes of Burgundy and Orleans was so great that much evil came out of it in England. There are many more: the Duke of Sommerset, uncle of King Edward, had his brother, who was an admiral, beheaded for having suspected him of wanting to encroach on the government that the Duke, his brother, had. Then, after the Duke of Mothombeland[148] overthrows the Duke of Sommerset, and he then seized the best part of the authority of the said Kingdom. There are enough others who are perished by the envy that has been brought to them. It seems to me to have proven well enough that envy must be held as an enemy.

Chapter XVI - How these terms "You do not know what you are doing" must be understood

These are words that are made quite commonly between gentlemen. Some say them without thinking about it. Others by presumption. And, in fact, these are not words that should be used between gentleman friends, because to say to his friend, "you do not know what you are saying," it is properly the denial of what he says, and one makes an enemy of his friend, or at least a dissatisfaction that one has of his friend. Because when the one says to someone he does not know what he is saying, it is as much as if one tells him you are indiscreet, ill, have no understanding, or else it is not true what you say and you lie. I will speak about a speech which happened at the Louvre at Paris. A Marshall of France, being in the chamber of Villequier where he took his dinner, the Marshall began to devise beneficial matters with very honest gentlemen while awaiting the said Villequier. One gentleman of the troupe spoke to him of the benefit which was near him, and of good value. He responded, "This benefit is not to my devotion, however, my good friend. You do not know what you say." This gentleman, aggrieved by that which was said to him, "Pardon, monsieur, I know well that which I say and it is not outside my understanding." The Marshall said to him, "You take it wrongly. I do not intend it from you. Pray, do not be angry for this, for I am one of your friends." From a gentleman to an equal gentleman, this could not pass without a quarrel and if it was one of whom you did not have much acquaintance,[149] and to whom you were not obliged to maintain re-

---

[148] Northumberland? The type is unclear.
[149] *pas grand cognoissance*

spect, and who said rudely to you, in a gallant and audacious manner, I believe that if one gave the lie, that he would be well for him to give the reason for the lie.

It is necessary that the other give him reasons for what he said. It is necessary at this time to speak of those who say, "You do not know what you are doing." I am of the opinion that this is not more properly spoken than of the other and, in this case, there would be as much motivation to give a lie. For to tell a gentleman rudely, "You don't know what you are doing," would be as much as calling him mad and insane. All those who use these terms are ill-advised in their language. It is to be given the lie of gaiety of heart, and it is proper motivation to draw the sword. I am assured that enough honest men will agree with what I say. I will say further that all doubts and misunderstood words, the brave chevalier must have it explained so that he remains content, fearing a reproach, and being accused of having no feeling. It seems to me that it is out of place to speak of these two terms because very often there are quarrels on this occasion.

Chapter XVII - Of those who tell themselves, "I am a good man and a gentleman of honour". How this must be understood.

I find that today this language is too much used, and even though it is, we must nevertheless look to whom we say it to, because between equals there is no need to speak of one's nobility, of one's valour, or one's honour. But it would be good if one said it to a gentleman of quality and good family, being offended, to tell him that "I am a gentleman of honour and a good man who would only want to endure [that which] you or anyone else would offend me [with]." We will speak more about it. No gentlemen can call himself a gentlemen of honour, because this word "honour" must serve the quality of the gentleman who has received the honour and ranks and who has command. He who is a gentleman of honour, if he were a gentleman of good and old house, who lived honourably, well esteemed, and greatly allied, he would be a gentleman of honour. But I am talking about gentlemen of the middle order and who usually frequent each other. That is how I am of the opinion that, regardless, all cannot be said to be gentlemen of honour. I will bring back to this subject an honest remonstrance made by a gentleman of quality and house to another gentleman, his neighbour, who having some business together, said that he was a gentleman of honour and a gentleman of means. It is an honest gentleman to whom he said these words, and not knowing the occasion which moved him to say it to him so often, he replies to him, "Never take this title that you have not deserved, and when you have deserved [it], or being honoured with some rank, you will call yourself a gentleman of honour. But in this hour, it must be enough for you to say, I am a gentleman and a good man." This gentleman thanked him for the honest remonstrance that he made him, and that, for his honour, he would retain it. When I speak of a gentleman of honour, I do not mean to speak only of those who have a lot of income, but I mean also of those who are of good stock and who live honestly with an honest reputation, being much esteemed for their good advice. These are worthy of being esteemed. To explain these terms well, it is necessary to know that all those who take the word to themselves, and that one says, "I am a very good man," is to honestly offer a fight to his companion. Example: here are two who are having angry words[150] and one says to his companion, "What do you mean? Do you have something in your heart that bothers you? I am a very good man." Saying this, it seems that he is presenting himself to fight if the other wants it. I will speak more fully of the gentleman of honour, and say, that all the quarrels which take place between chevaliers, they are made for honour. To achieve this point of honour, it is necessary to take time and age in order to govern this honour that we propose to follow. And like those who want to learn the sciences, and confine their lives to be learned and well experienced, they must use a lot of time and make a good foundation from the beginning of their youth to become learned. Also, of the gentleman in his tender youth, the principles must be well founded before carrying the sword so that when he comes to the age in strength, he can debate his honour, and make himself so experienced and capable that with strength and courage he can debate his honour with the sword. This is how the gentleman today takes this title of brave chevalier. And in truth, this degree of chivalry has been ordained by brave and bold persons in order to recognise from what depends on their honour and the means to defend [it], which the honourable chevalier must conduct well and as the laws of chivalry command and, if he proceeds otherwise, he will acquire from it the reputation of a bad chevalier, and that he also takes care not to make quarrels which would be more harmful than honourable to him. This is why I would like the gentleman to make such a good foundation in his youth, like having to go to foreign countries,

---

[150] *en picquer de parolles*

then knowing and frequenting great companions who would shape him with age, who would lead him to know himself and to execute beautiful designs from which he would receive contentment and honour, and then he could be named gentleman and chevalier of honour.

Chapter XVIII[151] – If calling another angry must be taken as an injury

Anger is an evil that possesses all men, and if it is not moderated, I shall properly call it rage, especially since the man, unable to limit his passions nor the pain by which he is most often agitated, is very often transported outside of himself. This anger is vicious and unbearable because those who are possessed of it, on very slight occasion offending their friends, cannot control themselves. This is the reason why it is placed in the rank of vices.

On the contrary, it is highly commendable to see a gentleman, furnished with a good modesty and a wise countenance. Those who are thus perfected are greatly to be esteemed, and they must be sought. This perfection proceeds in part from the nature of men. However, in one or in others, it cannot be but there is some anger. But, in one, it is more modest and, in others, more furious and vexing. For I know of no man, when he is offended in his honour, who may not be pushed to anger. But there are some who know how better to step than others, and say that those who are quick to respond,[152] and with a spirited promptness violently repel the injury that has been done to them, those are of great value, and in those one should be proud. But those who dissemble in order to feel it in time and place, it is necessary to beware of it, and in fact to tell the truth. It is properly disloyal to a gentleman to call him angry because when men neither reason nor judgement they approach the nature of beasts. Now a passionate anger which cannot be commanded or put away by reason, is worse than a brute beast. Even the melancholic anger, which is cruel and vindictive in nature, for the extreme and violent passions which overcome him, he deploys in his mind all strengths of vengeance and cruelty to satiate his pain. Seneca teaches distancing oneself from all cruelty and from the anger, which is minister of cruelty. The Frenchman, who cannot moderate his passions, is tainted by this anger because he lives in the region of Midi,[153] where men share in melancholic anger, and

---

[151] Incorrectly labelled Chapter 17 in the original.
[152] *repartir*
[153] the South of France

these are not easy to appease. It is the reason that those who are in this mood become furious and insentient. However, there is a difference between furious and insensate anger. Melancholic anger is the wisest, and when it becomes furious, its illness is more difficult to cure. And to a furious man, a curative is given to him. For insensate who are sanguine, no curative is given to them because, properly speaking, an insensate is he who cannot command his unbridled passion. I have wanted to talk about anger up to this point, so that we know more clearly what is called anger. I will say therefore that one kind is held to be an injury, namely, when it is accompanied by fury; the other is courageous anger, full of all valour. Those, when they know how to command themselves well, they are more praiseworthy, and when they want to execute their intentions they are prompt and diligent, without any reproach, not wanting to make any act that is ugly, nor blameworthy, to them and to their posterity. Seneca is of this opinion that severity is the closest form that may be to justice and, if it results from anger, it is vicious. There are men who are so often overcome by envy and a hatred that they cannot be reconciled and it is not in their power to moderate the passion which torments them. This proceeds from a hard heart which cannot be put to an end by reason. I therefore will say that the passionate and furious angry man is full of vice and, being vicious, one must avoid his conversation, anger being considered a vice. He who is called angry must consider himself injured. This consequence is valid.

Chapter XIX - How these words should be taken, "Take it as you will"

Those who feel offended by word or deed are advised to draw an honest satisfaction from it, and if it can be done amicably and to the satisfaction of the parties, it is much more honest. But, there are some who are so out of reason that they take no advice from their friends in their quarrel, and do not want to submit themselves to any satisfaction, so arrogant and uneasy are they to embrace reason, and very often leave their arbitration without doing anything. And after their arbitrators remonstrate with them a great deal, one replies that he did not care, and that their enemy should take it as he pleased. Another, when they have asked someone for a word and knew they could not draw from them the satisfaction they hoped for, they are embittered by it. This is the occasion that they respond to them, "I cannot say anything else to you. If that does not satisfy you, take it as you wish."

They ask how these words must be taken and understood. I answer that it is obviously to deliver the fight to the one who negotiates with his party, and that it is as much as if he were saying to him, "Since you do not want to take in good part what I tell you and you do not want to be content with it, I blame you,[154] and take it as you wish. For I will never give you any other reason nor less of an excuse." Here is what I have to say about these words which are proposed in this question. If he to whom this language is said does not draw his sword, he does himself wrong.

Chapter XX - Of the fear that the chevalier might have of his enemy

I am of opinion that there is no man so valiant nor so well versed in arms who does not in some way fear his enemy – except some rash ones who have no judgement and reason. And so that the valiant men will not find themselves scandalized by this opinion, I want to take the trouble to clear up this question. Firstly, I will say that he who fears his enemy should not be considered a coward for this. If it were that he had such fear that it turns [him] to flight,[155] then he should be rejected from the company of valiant men. But I don't intend to speak about flight in fear. I only speak about fear. Also, one must be so wise that when one has a quarrel with a valiant, courageous and well determined man, if one were challenged to have no fear which hinders [him] then one has not the heart and the strength to resist it. But the real fear that one could have of his enemy is that he will take you to his advantage, or prodigiously offend you. But when dealing with a brave chevalier and valiant gentleman who would not assault his party to his advantage, one must cast aside all fear. For this reason, a valiant man will never fear his enemy. I will therefore say that to fear properly in matters of quarrels is to be on guard and to guard against being surprised by one's enemy. This is not vicious, and to him who is not in a good condition for it, I would hold him as a man of little judgement and without conduct, as I said above. To those who proceed in this manner, it has never ended well and they are always overcome in their quarrel. I well know that there are some who will find my words bad, and will say that there is never fear without flight. To resolve this question, I will help myself with the ancient and wise

---

[154] *ie vous en met au pis*
[155] *si ce n'estoit qu'il eust vne telle crainte qu'elle tournast en peur*

Philosophers, who when they wanted to speak of fear, have put it into two species, and have figured one good which concerned the maintenance of republics, the other bad which was at all devoid of beautiful and laudable reasons, and, where there was no strength nor valour, those were considered pusillanimous. These wise philosophers said that not to fear anything was very harmful, but to be afraid was to prepare and fortify oneself to better defend against it. Plutarch greatly esteems this fear, and considers it a virtue when it is good, saying that it is necessary for those who have the authority to command, because those always fear and are afraid of doing wrong. Alexander said that there was no place, neither city, nor castle so strong that could reassure a fearful man. It is therefore necessary that the gentleman of honour never be afraid nor fear his enemy that this fear is not vicious, but which he may be filled with a brave and bold courage. Here is the fear of which I intend to speak, which is very virtuous to a man of valour, and if he is not accompanied by this beautiful virtue, I take him for a man of nothing and without judgement. I have heard it said that in the fights which take place today that the one who has the advantage over his companion makes him surrender his arms, and the other if he is but a little wounded surrenders. Here is a very shameful fight, and I cannot think that it is not for lack of courage, because however much he is wounded, he must always fight his honour until death, and would rather be killed than surrender, if it were not that the sword had come out of the hand and that his enemy gathered it up. But to surrender voluntarily is not done by the chevalier of honour, and if his enemy wanted to take away his weapons by force, he must resist until the last breath of his life, and tell him, "I will not surrender my weapons, kill me rather." Also it seems to me that he who would have an advantage over his enemy should be content with it without seeking this glory, because the renown[156] he has obtained from his enemy are quite apparent since he has wounded him, which is something that cannot be hidden. And it might happen that some brave chevalier will have to deal with him, who would do the like to him. For to a valiant man putting to mind another valiant man, it is to make him think of his conscience and his duty. But if he has business with a man of little means, he will disdain him and will not care,[157] like a mastiff to a small dog. But a valiant chevalier, bold & courageous, does not measure himself so easily. You have to think carefully when you go into proofs of arms with him and forget nothing of your duty. The valiant man will sooner choose death than to shamefully undo

---

[156] *marques*
[157] *s'en battra les ioües*

his honour, and then when he surrenders his arms he is undoing his honour and afterwards it is very difficult for him to recover it. It is true that if the fight were to be decided in the lists, he who would have obtained the victory over his enemy would be victorious, both in arms as in the body of his enemy. But when the fight is done by a head to head challenge[158] and promptly executed, it must suffice for him who has wounded his enemy to stand victorious. I advise all gentlemen of honour when they enter combat that they be desirous of preserving him well, and of knowing how to conduct himself so well that nothing can be reproached to them for falling into any infamy. For when one lays down one's arms, one must pray to God nobly and no longer carry a sword at his side.

Chapter XXI - That the weapons that chevaliers today present are not reasonable or in use

The weapons that the chevaliers present in their fight have never been used, and one cannot know who was the first inventor of them.[159] Truly, they are advantageous, to strike from further away, and the dagger likewise is rendered advantageous with a well covered shell. But this custom of combat has never been practised, and I believe that there is no more legitimate combat than the sword which we have been accustomed to carry on our side, and of which we understand avoiding the difference might befall us. These weapons with which one fights today cannot be carried easily. It is almost a mule load. I see some who carry them on horseback instead of their accustomed sword. I have seen two battles at Rome, among others, one in which the two parties fought with sword alone by the leave of the captain from the Castel Sant'Angelo, because they were his soldiers. And the captain said to me that if one of the two had carried with him other weapons than his sword he would have stabbed him with the pikes, even though he would not have said so, because it is the soldier's life to settle his quarrel with his sword. If this was not a fight which was granted to be finished in the lists, the one who is called there has the selection of arms according to the rules of duelling. I therefore opine that if he to whom his arms were presented, returned them to his enemy, he would do no fault or harm to his honour, offering, however,

---

[158] *vn appel cap à cap*
[159] Ed: I have no idea what weapons he is discussing here. Is he talking about pistols?

to be at the place which was assigned to him and to fight him with the sword he is accustomed to wear, and to show it to the one who came to call him. Since one must fight in a doublet, one must finish this fight with the sword we usually wear, and if the one who has called out his enemy refuses him, he must be judged as vanquished. I believe, however, that he will not do it. Also, he who calls out his enemy and presents to him such weapons must put [it] to his choice or take that which he presents, or fight with his accustomed sword. It is therefore necessary to settle the quarrel with its accustomed weapons. This is the duty of the chevaliers.

Chapter XXII - If someone, having bought a horse, is recognised and vouched for by an army, if he must return it

One can vouch for and make an arrest of a thing that one has lost and that one recognizes as one's own, and to what extent that the one who bought a horse has bought it without fraud, and has paid for it well. He could in any case be forced to return it, being unable to bring forth its seller. But in time of war, and even in an army, if the horse has been bought even though it was authorised, I am of the opinion that it should not be returned, given that it was bought in good faith and in an army being in the service of the King. It is an opinion that must be debated before the Captains who must maintain the right of the one who bought the said horse, otherwise it would be madness, without having any more means of being able to serve. It is something other than the law of war and civil rights. The duty of war must be maintained and preserved for the soldier, either for his honour or for his arms. And when he cannot find the one who sold it to him, it is enough for him to demonstrate by testimony that he bought it and paid well. But what can he who lost the horse say? I am in the army. I should not lose my money any more than he should lose his money. I would think that if he offers to wait for the money, he should be obliged to take it in order that he has the means to buy another. On this opinion I will put here a quarrel which happened for a similar reason to two Gentlemen at Lusignan being the Royal army with the third company, one of whom recognised a horse that the other had bought, which he did not want to return. They wanted to believe their *different*[160] Monsieur le Marquis de Villars, who had since

---

[160] Ed: I cannot figure this sense out in context

been made Admiral, could not make them agree, in taking the advice of the wisest Captains who thought that the one who had bought the horse should not lose it, at least not his money, given the time and the necessity of the war in which we were, on which, he could not agree. But on the strength of remonstrances, he was condemned that he would pay half of what the horse was worth, which Monsieur the Marquis de Villars paid at his own expense to avoid the dispute, which was thirty five *ecus*, without wanting to be reimbursed. It seems to me that my opinion is well proven by the advice of many honest gentlemen, all Captains. Let us see at this time, if a horse taken against the enemy in time of war is of good prize? It is necessary to specify this article. I say if a horse is taken going to war, that the horse, the soldier and the arms are well taken, and the soldier must be ransomed, regardless that he had borrowed the horse and the arms. Also, is it seemly in assaulting a town that all that which the soldier takes is a good prize? And it cannot nor must not be asked for nor vouched for by anyone. But if it is taken from his house by night, without the guidance of a Captain, it is badly taken. And he to whom such an act has been done could hold that this is an act of thievery. The Romans wanted thieves and robbers to be hanged and put in the gibbet as an example. With all the more reason the murderers should be cruelly punished, and also those who scale houses at night and break down the walls. These deserve greater punishment than those who commit their theft by day. The Muscovites and the Tartars condemn the thief to death. In the East Indies, in former times, thieves surprised in their larceny were impaled alive. But in the republic of the ancient Egyptians thievery was commendable, and the most subtle thief was the most esteemed and the most prudent, and although the Kings of their country had made several laws to correct this vice, it nevertheless remained eternal in their nations. On this question that I have made that a vouched-for horse should be surrendered. I came across this subject of thievery that I have wanted to speak on, it being of fairly specific subject matter, and also that the best police force that is in an army is to put order to thieves and murderers, which is the place where it is done the most. The military police are greatly needed for the safety of the soldiers, and of all those who follow them.

Chapter XXIII - A Captain who loaned horses and weapons to the soldier, should they be paid for?

I am of the opinion that the Captain who has provided a soldier with weapons and horses, by obligation, should be paid at the first showing,[161] at the price which will have been agreed between them. The showing done, the soldier, having been obligated, is bound to pay it. And even though he did not pay it precisely at the showing, it is enough that he received the money. But if it were thus that the equipment had been lost in the war, or that the soldier had been killed there without making any showing, I believe that the Captain cannot ask for his money, and the obligation must be lost and stand at null. But if it were that the soldier had kept the equipment that the Captain had sold to him more than one or two years, although he had not made the showing, I am of the opinion that he should be required to pay at the price which he was obliged to, because, since he kept it, by that, the fact goes to to prove that he wanted to use them. Therefore, since he uses it, he must pay for it. And even if the horse was dead or that he had gotten rid of it, that does not prevent him from still being obliged to pay it, for two reasons. One, because they used them for a long time after their journey. The other, that having rid himself of it, he, as if it belonged to him, disposed of it, and derived profit from it. But I would find it good that on returning from the journey, if he had not made a showing, that the soldier went to his Captain to ask him to take back the equipment that he that sold to him, provided that it was in its entirety. I would venture that the Captain would be bound to take it back, and return his obligation to him.

Chapter XXIIII - That he is not to look at his friend's letters

When we want to give certain of our friends our news, we write to them, and send them some of the business that we may have together, or of what is said and happening in the company where we are. In this way, the letter written and signed by our hand is a proper testimony to make our friend certain of the friendship that we have for him.

---

[161] *à la premiere monstrè*

That is why one must be as courteous[162] with one's friend as when meeting a messenger who brings him letters to not look at them, and doing otherwise, he is performing the office of a bad friend. For in looking at the letters that one has written to his friend, we offend two: he who writes, and the one to whom the letter is addressed. Also, doing such an act, it is to rush into a big and very rough quarrel that one could not appease as one would wish. Alexander the Great once received a packet which was of very great consequence. He retired to his room to look at it. Hephaestion, whom he loved singularly, took one of the letters which were in this packet and looked at it. Alexander, not wanting him to cause displeasure, had a ring on his finger, and put it in his mouth, meaning by this that Hephaestion must have two things in recommendation: one, that he did not read these letters aloud; other, after the greetings, that he recited what was contained therein. One should not therefore look at his friend's letters, so that he has no occasion to complain that he has been made a bad friend, and also to avoid a quarrel. For this fact depends on fidelity and virtue[163] of a good man, and of an honest Gentleman.

Chapter XXV - It is very honest for gentlemen to salute each other[164]

It is a mark of recognition which has been observed for all time with honest gentlemen to salute each other, and those who do not want to use it, and who expect to be saluted first, one holds them for enemies, and very little bred in civilization. Plutarch says that it was the custom when one meets his enemy to cover the head, and before his Prince and his friend to uncover himself. For the head is the most principal member of man, and the most dignified. Also by saluting him one puts oneself in his power, calling himself his inferior. Also, in truth, it is a sign of honour and reverence, when one is uncovered and humbles oneself. We have the very worthy example of Fabius Gurges, a young man being consul, seeing his father come to the Senate mounted on his horse, asked an usher to warn him to dismount, which he found was very good, and did honour to his son to have understood his charge well and knew how to maintain himself well in the ranks and honour to which he was called. This example is not far from reason, since when one is called to some title of honour

---

[162] *officieux*
[163] *preud'hommie*
[164] *de se saluër* To greet each other, to do courtesy to each other

one must make a difference with those who are not of similar quality. Formerly in Rome to make the commoner different from the nobility, who were ordinarily in controversy and abuse, it was advised that the commoners would make Tribunes and that the nobles would not be received there, and the Consulate would be given to gentlemen without commoners being able to aspire to it. In such a way that the loss of great honours and consulates was only open to the nobility and the commoners hardly ever appeared there, unless it was for having done great services to the republic and many acts of war. Of note: like Marius who from a peasant became such a gentleman and had seven times the status of Consul. Yet this was done with great difficulty, because there were only nobles and families of very old houses who possessed these ranks. So it is very fitting to the gentleman by his virtue and his valour to acquire great honours and great charges like Bertrand du Guesclin, who was called by King Charles V to be his Constable and did him great services, and many others that by their great virtue and magnanimity have been bestowed with the highest degree of honour that can be desired. Since, therefore, valour and long service are the occasion that one is bestowed by the King with honourable offices, it is very reasonable to honour them and salute them. I have well wanted to note this example to fortify the honour and the reverence that one owes to those who have qualities and in which honour formerly they were held.

Fourth Part

In the third part, we have spoken enough of the arbitrators and of their conditions, whether it is necessary to know their names and, similarly, of those who mediate the reconciliation of a quarrel. Also, as a gentleman must keep his promise, so governors of provinces should take cognizance of quarrels. Likewise, a prisoner of war should keep the oath he has promised. We have also reported the difference of a gentleman who claims to be of a better house than another, and that the gentleman of honour must be careful not to be afraid or show fear when he enters into combat, along with several other questions and requests which are proper to infer on this subject. We will speak at this time in this fourth part, after having put the Chevaliers through the fights, and having debated their honour, and done all things that come from that, of the most honest means that the gentleman can take to avoid quarrels.

Chapter One – The civil and internal wars are partly the cause of the abundance of quarrels

After having spoken of the several benefits of quarrels, and of the reasons which move gentlemen to quarrel, together the manner of resolving them, it seems to me to be good to add to them that from which quarrels proceed. I have always considered that the abundance of quarrels and their main impetus results from civil and internal wars, which are made and composed from the diversity of the opinions of the victims who enter into differences, or from the envies which are generated between the greatest. From there, they undertake to rise up. Spite makes them revolt and take up arms under the pretext of some sudden incident (especially for the Frenchman, who is quicker to do so than any other nation). All are of the opinion that civil wars have been the ruin of kingdoms and monarchies, and also of republics. The histories give enough testimony of my words. We know enough how the flourishing republics of Athens and Lacedemona were constructed and built by the greatest legislators and warriors of their time, who did not know how to avoid the peril of falling into dissension, which Lycurgus and Solon foresaw, two of the greatest masters to properly govern a republic. They knew so well how to put republics into order that they made them last a long time, and yet they did not know how to last so long that they did not make themselves prey to their neighbours, each carrying off his share. The Romans, who had been so triumphant and so bravely victorious, through dissension among the greatest, fell into bloody civil and internal wars, that in the end the strongest prevailed and made of it an empire which reigned for a long time. But as long as they had something to discuss with their neighbours, they did so with the happy success and preservation of very great captains. This was the reason when Scipio had the country razed to the ground that he exclaimed, "Ha! Rome, here is your loss, and of all the republic", and in fact razed Carthage to remove the exercise of the war from the great captains of Rome and from the youth who entered the civil wars immediately, like Sulla and Marius, Caesar and Pompey, Augustus and Marc Anthony. Finally Augustus Caesar remained master of it. At that time. We only saw quarrels in Rome, some taking one side, and the others the other, and from there proceeded many murders. And from which side did we see the Guelphs and Ghibellines, the seven holding the party for the Empire, and the others for the Pope, and the partisan lords were fighting each other at war. On this occasion, it would be a confusion to recount all the quarrels which are fought on the occasion of civil wars. Holy and prophetic histories are full of them. It is found in the histories of France that the quarrel between the house of Burgundy and Orleans lasted nearly fifty years, during which great murders took place on both sides, which was a pitiful spectacle for France. Thus, the subjects, being of opposite parties, enter into quarrels for complaints or other excesses which this miserable war produced, such as the ransacking of houses, imprisonments, extraordinary ransoms, murders, and, whoever had quarrels, it was the instant and the time for revenge. Here is how it is apparent that the civil wars have produced the quarrels which reign today. I am not capable enough to give a remedy which would be proper to appease these quarrels. It would be expedient that a wise legislator was called to do it. The King who is reigning today with his prudence will put forth such good order there that all his subjects will remain satisfied to keep themselves in peace, unity and harmony. Aristotle proposes several reasons that disturb a state. I am not here to reproduce them because the style is too long. To those who are curious to see them I would send to reading about them. They will find there good reasoning, all apparent and very considered, to maintain a state in its greatness. To avoid this confusion, I judge that it is necessary to exercise the subject of a war with the foreign neighbour, not to continue it [the internal strife?]. That one looks at as much that we had war with a foreigner, if we saw quarrels in this Kingdom. They were so rare there that they were hardly mentioned, and if there were any, they were continuously appeased since we had peace in France with our neighbours, a year later we fell into the civil wars. Thus it seems to me to have well proven that civil wars are the cause of the abundance of quarrels.

Chapter II – That the surest means of avoiding quarrels is maintaining piety. Justice is the true foundation of maintaining concord and friendship.

We have spoken in the previous chapter of the cause for quarrels. It seems to me that it will be very appropriate to speak at this time of the means to avoid them. The means which seems proper to me is maintaining oneself with piety and obeying justice, being fully certain that these two columns of concord are maintained with all the honour and obedience that one must give to his King, with the goodwill which must be undertaken in society and in dealings with men,[165] and I believe that there is no man of sound judgement and knowing that this is the reason who does not embrace my words.

I will speak of that which I find most evident and worthy of being well considered. I will start with piety. Piety is taken as one who is good-natured,[166] humaine, charitable, peaceful, full of devotion, temperate, and full of all good parts that a man of virtue should possess. I would say therefore that he who is merciful[167] and full of goodness is furnished with great friendliness and very quick to please his friend. Those who are furnished with this piety are peaceful, hating

---

[165] *frequentation des hommes*
[166] *debonnaire*
[167] *pitoyable*

dissension and quarrels, not wanting to offend a single person, fearing to have arguments with anyone, as they are full of probity[168] and goodness. The conversation of such persons is to be greatly desired as, on the contrary, that of these furious and quarrelsome fools, who are always inclined to evil, is to be avoided, and because it is a very difficult thing for each of them to be able to repress these emotions, to contain the movements of his heart, and to moderate the way of his life, and to maintain himself so happily that nothing can make him fall into something dishonest or reproachable. This should make us think more closely about forcing us to look carefully at our actions, so that we do not undertake anything that may detract from the reputation that the virtuous man should acquire. By these honest and holy circumstances, we will avoid quarrels. For any peaceful and well respected man will always have the occasion for a dispute, but will not stop for words which are light and of little impact, as the turbulent usually do, who take offense at all things, without any appearance of reason. One must flee from these people who are quick in words, and watch what they say, and who do not know when they speak whether there is virtue or merit. And in fact one cannot make a distinction between the wise and the fool, except that the wise man, by his good words and judgement, always seeks virtue, and conducts himself in his affairs with complete fairness. But he who is a fool is always driven by his own sensuality, using his passions in his own affairs. From this, it is easy to judge how much the man, full of truth and who follows it, is happier than the one who is full of madness and lies. The truth, therefore, is desirable because it makes man shine in all his acts and words in such a way that each one is affectionate and very eager to seek the company[169] of such persons, for from them you will only learn good works, and to hate and to detest lies. These people should be frequented because they are full of virtue, and their lives filled with wisdom, goodness and modesty. With these people there will be brewed no quarrel, and they will advise you to flee them, as being a total ruin in a troop of men of honour. So, we must separate the quarrelsome from the company of men of honour and virtue, as a mangy sheep is always separated from the herd of those who are very healthy, for fear that this one by itself infects the others. When a swaggering quarrelsome man[170] has long frequented a company of men of honour, he reduces them from better conditions. This is why good company is to be desired and the evil ones must be avoided. With

---

[168] *prud'homme*
[169] *frequentation*
[170] *vn querelleux fort euenté*

good people, one only learns goodness and honour and to blame vice. With the vicious, one only sees wickedness, in which we most often allow ourselves to be carried away by lack of judgement. Bad company causes loss and dishonours the gentleman who wants to obey their designs. And from there, it happens that the house and the family from which he is born, receives shame from it. There is therefore nothing so necessary to the gentleman in order to avoid quarrels than the company of virtuous men. This is what I have to say about piety and the truth, joined together, and observing them well is the means to obviate quarrels. And he who scrupulously keeps them will make himself happy and well accomplished in all honest exercise of virtue, and will avoid the many quarrels which are made every day. We will speak later in this chapter of justice. But humility must not be left behind, which follows piety closely and is proper for the gentleman, with which he must accompany himself in order to be the perfection of honour. And he will have difficulty frequenting the company of the ranked and honourable if he is not humble and gracious. It is from this that he must derive a laudable reputation to be loved, esteemed, well wanted, and cherished. On the contrary, if he is arrogant and full of glory, he will be odious and neglected by everyone, from which will ensue a lot of quarrels, and several other inconveniences. Very few gallant men can sympathize with a conceited one, and these are hated by each one. It is also the way to keep in a good concord and union because, if between relatives, friends and neighbours there is no concord, it will be difficult in time not to bring some trouble to it, given that the infirmity of this human life which is subjected to so many accidents, seeing the truth of a thing, which at every hour changes and diminishes.

We must consider magnanimous and virtuous those who know how to resist it. This can only be accomplished well by this virtue of piety, and cannot be acquired by the nature of man alone, but only divinely, and by the good will of God. He, therefore, must be considered wise, magnanimous and virtuous, who by such holy consideration and close friendship, maintains and preserves himself with relatives, neighbours and friends.

Chapter III – Of Justice which is companion of Piety

Let us speak at this time about the justice that I had reserved to write about in this chapter. It must be maintained and preserved, as much for the good as for those who are of a bad life. Justice is nothing else than rendering duty to everyone equally, with all reason and equity. Aristotle approves of these words when he says that, "no one can call himself just if he is not accompanied by the good will to do everything justly." Therefore, one values he who lives according to the dictates of justice, and who fears change in things which are equitable. He will be governed by God because in him is the wisdom of justice which leads all his thoughts and actions to do good. This covenant which he has with God makes him praiseworthy among men because he exercises all honest charity, and makes himself liberal and benevolent to his friends. Contentment in human life is to do well, to be praised, prized and rewarded, both in life and after death, and to immortalise forever the memory of his greatness for posterity. The wicked leave a memory of their actions to their posterity as well. But this is not with virtue nor with works that are praiseworthy, like the tyrants[171] who exercise some kind of justice. I will report in this place a very cruel example of Cambyses, who had Sisamnes flayed, who had unjustly exercised his justice. He had the skin glued to the judicial bench.[172] How much of this work was done for good reason if nevertheless it was done rather to satisfy his accustomed way of tyrannising people, than for a good desire that he may have had to do justice. The wicked are never furnished with good reason and have no justice in them. But, the good man who is assisted by the fear of God will always embrace virtue with total honour, and will maintain himself with his friends with full duty and equity, being endowed with the divines graces with which God has adorned him, which are the mirror of all excellence, especially when it comes from the total power of God. It is therefore with good reason that I maintain that the support of justice makes us maintain [ourselves] in friendship and concord and prevents a multitude of quarrels. Certainly, there is nothing found which has greater strength to contain he who is haughty, turbulent and quarrelsome than the fear of God and the reverence which he must carry for justice in obeying his King and the magistrates. This is the bridle which stays the vicious man and prevents him from abandoning his life and goods to perpetual dishonour for his family and reputation. I am nevertheless of the opinion that it is very, very difficult for one to be able to remove his mind from vice, who will not apply himself to properly contemplating the renown of virtuous men and how admirable it is. And he who wants to carefully search for the excellence which comes from it will make himself honoured. On this subject,

---

[171] *Tirans?*
[172] *la chaire de iudicature*

the ancient philosopher Chrisippus, to demonstrate how much justice must be admired among men, had a virgin painted with ardent and sparkling eyes in order to prove that justice must be guarded inviolably. With all fairness, Lysander, Captain of the Lacedaemonians, was of the opinion that the best governed republic was that where the good and the wicked were compensated according to their merits and faults. The Pagans held justice in such reverence that they honoured after their King, the magistrates who exercises it, and he who was so reckless and brazen as to touch a magistrate was punished with death. Good laws and virtuous Princes make their subjects good and very obedient. And from well commanded order[173] proceeds the continuation and support of a flourishing state, and instructions to others who rule after him to follow this track. It will be sufficiently proven that piety and justice are the two true foundations for maintaining oneself in concord and friendship, which is the true means for a gentleman to avoid quarrels.

Chapter IIII – That the Nobility must be nourished in every honest exercise and learn that it is from virtue that one lives happily

Socrates, Plato, Xenophon, Plutarch, and several others who wrote about discipline have carefully recommended the good instruction of children. And they are of this opinion that youth will have been well instructed with honest discipline and that it will be felt when they have attained the age of majority. These philosophers also greatly blame the fathers who through avarice do not want to have their children instructed well. This is the reason that fathers are anxious to bestow on them a good start and to teach them to serve God and to flee vice. Be certain that if in their tender youth, they begin to do well, they will continue, and will not have a single desire for evil doing for the rest of their lives. I would very willingly blame most of the fathers who have so little solicitude for instruction of the their children, who do not want to take any expense for having any tutors to teach them. And when they have reached the age of twelve or fifteen, they do not know how to read or write. So often it happens that when their children have achieved the age of knowledge, they detest the bad principles of their adolescence, and feel ill-kept by fathers who have not

---

[173] The original has "police", meaning at this time the order of the community more so than the body of officers who enforce it.

taken care to have them better educated. They are so uncivil. They have no grace or comportment, and cannot say a word properly. This commonly comes from the ugly food they have been given. I speak as much about the rich as of the poor, for the rich gentleman, however great his faculties, would prefer to use his means to live honourably according to his Nobility, and keep large teams of dogs, birds, and horses, making an excessive expense on his household, than to regulate it and save [money] to have his children brought up and taught good letters, being a thing greatly [more] acceptable to a gentleman than letters and arms. All the others who make a profession of science, and who have not come from nobility, if they study, they do so to make use of it for the rest of their lives, and thereby they acquire means for their children, and for their posterity. But when the gentleman has knowledge, it is not to enrich his household with it, it is to derive some pleasure and contentment from it, and to make use of it when a necessity arises – all the more so as it is a rare thing to see a knowing and valiant gentleman who is a fine and honourable mark for his house, and for his country, and who will make himself some day capable and worthy of being employed in great negotiations. It is the way by which virtuous personages advance to aspire to an immortal praise. The ancient Romans instructed their children in letters, and prized no one in their republic who was not full of knowledge. They had such great reverence for gentlemen of knowledge, that they greatly admired them, and those who were not of such quality, if he knew them learned for the virtues which were in them, made them and their family gentlemen. Cicero testifies to it, speaking of himself, that for his knowledge and his eloquence, he was made consul and ennobled with all his family, and afterwards he did great services for the republic, so the letters report. There are sufficient of those who blame the gentleman who has studied, and say that his sword smells of the writing desk, a saying more proper to a man who knows nothing and who is ignorant of all. For the man of virtue will prize and honour the gentleman of knowledge, and will hold him very excellent. It is also a great shame to blame virtue and have it be a vice. Alexander the Great was instructed from his youth in good letters and had for his tutor Aristotle, and was recognised as a great and excellent monarch. King Francis I loved letters, and always had learned men around him, in whom he took great pleasure. So when the young gentleman has been taught the good sciences, he must prepare to go to Court to make himself known to the King, and if the King recognises in him some honest perfection, it cannot be that he will not take him to his service. It is also the place where he will find many honest men with whom he will make ordinary conversation, especially with those whom he finds more inclined to honest exercises. Here is what it seemed to me good to say for the virtuous perfection of nobility, and of the young gentleman. Plutarch has made a treatise on how children should be nourished, and says that there is nothing that is so effective in virtue and in making a man very happy as a good upbringing, and that all the others are small and very weak compared to him. He says moreover: although good food is very necessary, there are however limits[174] that must be followed to make the young man very perfect in virtue. We will speak of them in the following chapter.

Chapter V – Of the virtues that are proper to the gentleman to make him perfect and well accomplished

All the Philosophers have taught four kinds of virtues by which the mind of man can be instructed in an honest way of living well. The first is prudence; the second, magnanimity; the third, temperance; and the fourth, justice. I have spoken quite amply above of justice. Of magnanimity, we will speak of it in the following chapter. At this hour, I will speak of prudence and temperance, and first of all of prudence, which is a virtue so very laudable that no one is worthy to converse with Kings and great Princes unless he is accompanied by this virtue, because it supports and surrounds the man of honour with power and authority so well that it gives him the means to go everywhere. This valiant Roman captain, Fabius Maximus, by his prudence, lowered the fury of Hannibal of Carthage who had held the republic for so long in subjection. Agamemnon greatly esteemed the prudence of his Nestor, from whom he usually took advice, knowing him to be prudent and that he did not vary in his opinions. Also, it is necessary to be accompanied by people who are wise, modest and very discreet. The substance[175] of the prudent man is to examine his undertakings well and not to let himself be overcome by some false opinion, because, if you go beyond these bounds and limits, you will do many acts of wickedness and deceit, and everyone will hold you to be deceitful and the enemy of all reason. Humility accompanies prudence. The conceited man is hated by all. Pride makes the person dishonest, tyrannical, and almost barbaric, displeasing to God and to men.

---

[174] *circonstances*
[175] *propre*

Sacred Scripture sufficiently testifies how much God abhorred vainglory. I will speak at this time of temperance, and, firstly, of sobriety. The actions of the sober gentleman will accompany him, because he is always attired with an honest countenance, with the maintenance of body and mind, and understanding. It is quite certain that if the gentleman is excessive in his way of life that he will never do a good job and his entire vocation will be vicious. Enough great personages, for their drunkenness, have fallen into great misfortune. Solon, the wise legislator, made a law by which he commanded the slaying of anyone who was found drunk. Plato, as he arrived in Sicily, seeing the table of Dionysius the Tyrant covered with so many varieties of meat, said that such a service was more becoming to swine than to men. From this vice, there comes a most infamous thing, which is lust, about which I want to admonish the honest gentleman not to allow himself to be overcome by all these ugly affections which defile the soul and the body, ruin the mind and soul and the understanding, through which the women of great personages have been lost. What provides the occasion for this vice but sloth?[176] For the man who has no exercise is the one who sooner abandons himself to his overflowing lusts, especially since it is quite certain that the one who has no knowledge or education is idle and, being such, he applies himself more willingly to imperfections, because he has no perfection. This is the reason that the gentleman should be nurtured in all honest exercise of virtue. This will make him avoid vice and quarrels, and he will always exercise himself in doing some honest office in order to resist all the bad impressions which are customary to torment a man, and which only succeed in changing good opinions into bad ones. And being accustomed to practice something honest, his mind will learn to avert all the evils which could agitate him. This is the reason why I advise the gentleman to avoid sloth, which is the mother-nurse of all vices. It is also necessary, if the young gentleman wants to be constrained and well tempered, that he abstain from great expenses. This brings ruin to good houses. I know well that this heart of the nobility is so full of ambition that it cannot correct and regulate its expenditure. It is for lack of knowing how to recognise himself. For this reason, the nobleman must regulate his expenditure by good judgement and great providence, otherwise he will ensure that he will succumb and become poor. Wisdom and honesty maintain a gentleman. It is knowing how to measure oneself according to his ability. Thus, that which it seemed appropriate to me to write about the instruction of

---

[176] *l'oysiueté*

the virtuous gentleman, to make his life happy and to immortalise his memory when he becomes loved, to follow him well. He who desires to learn further, I would send him again to read Seneca, where he will see many good instructions which could be of use to good morals, for food (as Seneca says) and instruction mould us, ripens them, and each feels that which he has been taught. And for this, the good companion must take away what the bad one has introduced. I will therefore ask the young gentleman, if he takes pleasure in reading what I have just said, to draw an example from it in doing so. He will be happy who can help him to avoid the inconvenience of a quarrel.

Chapter VI – That the boldness and valour of a gentleman should not be esteemed if they are not accompanied by magnanimity

The magnanimous gentleman has always been greatly esteemed when he has boldness and courage, especially he who makes the profession of arms, and he who is most suitable for it, and who most often experiences it, and he who makes it shine in all places where magnanimity is found. Also, uneasily, the valour that one properly calls magnanimity, when it is accompanied by boldness cannot be hidden, which is a praiseworthy valour, prized by everyone. But vicious boldness, which is accompanied only by temerity and presumption, must be rejected by valiant men. Valiant men have always been prized. Indeed, the valour of a gentleman should be praised. It is this that greatly embellishes his fame, especially since he is prepared to take up arms. All that which he intends to do is in order to acquire the title of boldness, and if he knows that one does not does not hold him in this esteem, he strives to prove it by his valour, having a brave and generous heart. I therefore mean that the bold and valiant man, if he is not accompanied by this beautiful virtue of magnanimity, his value is very little commendable. For this reason, the valour of a gentleman must be of such condition that he does not make himself more haughty, nor more audacious nor reckless, but that he is accomplished with many honest people and manners, courteous, gracious, full of good morals, modest in his actions, sober, not boastful nor conceited, and not esteeming himself more than he should, except to fear being thus well accompanied by so many such virtues. The valiant gentleman will thus live freely and without conceit. Proper magnanimity does not vary in any way and resolutely awaits the end of one's life. There is nothing beautiful nor great in all human things unless it is a big heart and a bold courage, which embraces all beautiful and great things. If the gentleman is magnanimous and courageous, he will never think that anyone can do him wrong. When he sees his enemy in his power, he may consider it a very great revenge to have the power to avenge himself. The magnanimous gentleman will not attack his enemy until he has first made him understand, for cunning and deception do not live except in those who have a weak heart and are of little value. It is a beautiful thing to make oneself feared, and dreaded by one's enemies, and to make oneself loved by one's friends. The gentleman wears the sword to use it in a place of honour, and to repel the insults that one could do to him, and help his friends when he is required to do so. But he must be careful not to use it on all occasions, either good and bad, for in that, there would be no magnanimity, like many quarrelsome people who have no other exercise than to affront their neighbours. I know of good houses which have made themselves poor and needy for having maintained quarrels and having employed the best part of their means on them. The gentleman of virtue must follow this exercise. It is therefore necessary that the Gentleman abandon all his cowardly works which are of little merit, and that around Kings,[177] he seek the way to acquire the title of magnanimity by his valour and hardy courage. The King knows well how to select those he will know that have value and merit, and favour them. The greatness of a King is such that he will manifest magnanimity in all his generous actions, by elevating those of merit, and driving out those who are unworthy of him. Taking the example of Elagabalus,[178] who was so inhuman that he gave the most beautiful states of his kingdom to the most miserable villains he could meet. Also it is necessary that the virtuous Gentleman commit all his plans to do service to his King, and to stay close to him, so that some day he will be called by his Prince to serve him, and be employed for his province: sparing neither his life nor his property to give testimony of his Nobility and his hardy courage, both for his country and for his posterity. This is what I can say of the boldness and magnanimity of the gentleman, and how esteemed it must be.

---

[177] *qu'aupres des roys*
[178] Marcus Aurelius Antoninus "Elagabalus", born Sextus Varius Avitus Bassianus, c. 204 – 11/12 March 222), was Roman emperor from 218 to 222, and his reign is well not for religious and sexual scandals.

Chapter VII – That it is something that greatly weakens the boldness of a gentleman, who does not move from his house, and who does not seek the hazards of war

This question is worthy of being asked: whether there is something the gentleman lacks in the obligation of his honour, should it prompt him to take the path that a man of valour and courage should take? The most excellent exercise for a gentleman is war, which should be so violently imprinted on a noble heart that on all occasions that will present themselves, he must use his life and his property and, in doing this, his reputation will be greater and much esteemed. But when the Gentleman indulges in his pleasures, stagnating in his house with all the delicacies he can imagine, I cannot consider him very noble or full of courage, or similar to one who would expose himself to the hazards of war. However, in wishing to not judge him too rashly, I will make some separation of the one from the other. I will begin by speaking of the youth of a Gentleman, who being on the path on to bear arms, makes him more agreeable to the companies and more familiarly he seeks them. Also, when he abandons this first way, he forgets himself and practices something lighter, loving to take his pleasure in his house. It is misfortune for the gentleman that when he has once tasted and taken his ease, he can no longer devote himself to the exercise of war. However, this first start makes him a gallant, hardy and valiant man. Men, when they occupy themselves thus in their houses, do it for some necessities which are obvious, as having great lawsuits where they go most often with all their property and the others for debts. All that compels them to stay there to give instructions. To those, I give them an excuse. The others are in no way desirous of it, and are resolved not to move from their house for any occasion whatsoever. And nevertheless there is still some nobility in them, having come from a noble family. This is why they are naturally so inclined that it is not possible for them to do evil in their house. They most often devote themselves to many honest exercises, and with very honest and not too excessive companions. The gentleman living in this way, I highly esteem his way of life. But those are of a certain natural instinct with which they are gifted. Nature has not worked so with everyone, for there are those who are not furnished with either knowledge or understanding, and are useless in their homes and in all places to which one would like to call them, and have no other exercise except avarice and usury, so soft and weak are their hearts. Of these, I do not know what judgement I could make, and it is not my intention to scandalise them. But I can say with truth that the life of such a person is greatly to be despised, because there is no there is no exercise of honour in them, nothing to be commended. One will tell me that the exercise of war is very expensive, and that the poor gentleman cannot reach that far with his small means, that this forces him to remain in his house. I respond that in war there are enough ways for the poor gentlemen to seek his fortune. Princes, great lords and Gentlemen of good houses will be very glad to have around them gentlemen of honest ways. And if they want to be disposed to follow them, they will acquire goods there and will learn honour, and by this means they will see war which will cost them nothing. And by this means they will fashion themselves with such good grace, that they will be loved and respected by their kinsmen. I greatly deplore the gendarmerie of France, who I used to see so proper and so well maintained, and so dexterously ordered that there was nothing in France so admirable or more worthy of a gentleman. Also, this exercise was intended for the Gentleman, both for the rich as for the poor, and the parade of all France. I saw that when the order of the gendarmerie of France was well paid by neighbourhood, the gentleman maintained his place as a man of arms so much so that there was a great press to be enrolled in the company of gendarmes. But now it is so reduced and almost completely annihilated. Gentlemen have let themselves fall into all sorts of misfortune. I hope that he will hereafter resume his first habit, and then we will see the nobility flourish, and take a path different than it does now. I would like it to be placed in the same order to which King Francis I had returned it, who did not want anyone to be received into his companies of ordonnance unless he was a gentleman and knew the Captains, and that he would have at least reached the age of eighteen or twenty, and the leaders twenty-five. It was a fine order for the gentleman in those times, who stayed very little in his house, and needed by necessity to have good horses and good arms, to be ready to march for the services of the King. Also, King Henry II always kept it well maintained and well paid, and with the advice and prudence of his Constable, Anne de Montmorency,[179] in its perfection, it performed better. It was very superb and appalling to foreign nations. And in truth when a combat presented itself in which the French gendarmerie were involved, being well led, they

---

[179] Duke de Montmorency, b.1493, d.1567. A French noble, governor, royal favourite and Constable of France during the Italian Wars and the French Wars of Religion. He served under the French kings Louis XII, Francis I, Henri II, Francis II and Charles IX.

gave great defeats and more often won the victory. Today, the King begins to want to straighten it out. I very humbly beg his Majesty, as his very humble servant, for him to continue in such a holy work, and by doing so he will find himself very sure of his Nobility. In short, I must say that this is one of the supports and the maintenance of the Crown of France. Rightly, we must greatly praise King Charles VII who was the first who stood it up. King Louis XI, his son, instituted the Order of Saint Michel. King Charles IX had so much business that he made them more common than they used to be. He did me the honour of giving them to me, which I received from the hands of Monsieur de Montpensier. Then King Henry III gave me a company of men-at-arms after the death of Monsieur de la Trimoüille,[180] for whom I was a Lieutenant. It is not inappropriate, it seems to me, to recite in this place the honours which I have received from my Masters. But I am unhappy that I have done only a little service to do them whom I have always wanted to serve. To return to my first point: I am desirous that the gentleman practice war, and that he does not stay in his house, so that by this he acquires the reputation of being bold and courageous, and that he he does not allow himself to be buried in mechanical things, for that disgraces and renders the Gentleman notorious and of little longevity.

Chapter VIII – Following in this Chapter, proving that laziness in a Gentleman is to be avoided

Laziness is not proper, according to the nature of man. For this reason, the Gentleman of honour must avoid it. Our soul, which is never at rest, gives us very obvious testimony of this. And without applying himself to some honest exercise, especially since man is always inclined by himself to good works and, if he is discouraged by laziness, and for lack of good exercise, he will become master of many bad habits. Often the Gentleman licenses himself to do bad offices to satisfy his needs. This imperfection arises from where? It comes from failing to recognise virtue, and to embrace it strictly, for the virtuous Gentleman, if he knows he has a lack of possessions, he is very diligent and drives out all laziness in order to acquire them. This is how,

---

[180] Louis II de la Trémoille (29 September 1460 – 24 February 1525), also known as La Trimouille, was a French general. He served under Charles VIII, Louis XII and Francis I, and was killed in combat at the Battle of Pavia.

through lack of possessions, one can make for himself a great character, worthy and held in great honour. It is in this France where this proof has been made. That from a small house, we have seen them shine and grow by virtue and their swords have acquired for them more goods and honours with effort, and the time they employed in it, than the most miserly of them could have amassed being a long resident in his house. What difference is there between these two? The one has acquired goods and honour through his valour and the dexterity of body[181] and through understanding. The other by laziness and cowardice,[182] sleeping in the business of his house, having extended neither the heart nor care to anything other than keeping hold of *écus*.[183] And it only need that miserable war or some other misfortune happen for him to lose in one day all that he has acquired during his life. Or, on the contrary, he who has through his work and valour acquired wealth cannot lose them. And when he dies, his acts and virtuous actions will not perish, but he will leave his memory, which will remain immortal, forever in his posterity, having acquired good title and the reputation of a noble gentleman. Pythagoras approved of my words when he said, "That the difficult and most painful things will lead us sooner to virtue than the effeminate and pleasant." All as much to say, "That a man, through lack of heart, allows to pass that which is most useful and necessary, which is virtue, in order to fall asleep in laziness and to indulge in things of no value." This is the reason why the virtuous Gentleman must always prefer honour to profit. And whatever need he may have, let him not overwhelm himself[184] by doing things which are illicit, and which could defame his honour. One must greatly condemn a gentleman from a good and well allied house seeing him useless in the country, and consuming the best part of the day playing at skittles and cards and living a shameful life. From here proceeds an infinity misfortunes like blasphemies, quarrels and many other evils. What satisfaction do you think parents can take from this pitiful exercise? The whole family detests it and feels ashamed of it, as being a scandal to the house that this ill-mannered gentleman exercises without respect of the good people to whom he belongs, and of the house from which he came. Alphonse, King of Spain, who instituted the Knights of the Order of the Garter,[185] made

---

[181] The original is clear, *d'exterité de corps*, but this makes no sense. *Exterité* is not a word.
[182] *lascheté de courage*
[183] A gold coin
[184] *se desborde*
[185] Alphonso XI created this order in 1332

an ordinance that none of his knights should play cards or Tens[186] on pain of losing access to his château and to his person, together with their wages. Sufficient others do not move from their houses and serve only to accompany quarrels, make assemblies, and have no other advice but to beat one and kill the other. The value and the honour of a gentleman is not valued by forging quarrels, but I would rather believe that this would lessen his reputation. One should not consider him for that as more valiant but as a furious man without judgement, avoiding this deformity. It is more honourable to seek the dangers of war than to contain oneself in one's house with vice, and however much that from the first approach one does not arrive at the things to which one aspires, one must not for that reason distract oneself from his first thought. War must be pursued with such diligence that one can derive some fruit from it and strive to achieve some honest perfection. For those who aspire to high and great things, it is reasonable that they experience by several kinds of it, and of the most honest of which they are advised to bring back the memory of so many great personages who have taken to it with great honour, to end their days in some honest vocation. And to achieve this, one must seek the hazards of war and be considered a bold and valiant gentleman. Because the Prince will not send them the riches and the honours, especially to the house of the one who will never have done him service, and who has never left his house, it is therefore necessary to work to acquire this praise and avoid this vice of laziness.

Chapter IX – That there is more honour for the gentleman, when he is offended, to draw his justice from it with honest satisfaction than to avenge himself

The gentleman full of valour would not want to let something pass that touched his honour without trying to extract satisfaction for it by force of arms. And if he did otherwise, he would have to be understood as lacking heart. Truly, the honour and the boldness of the gentleman, with the shame that can come from it, is the main occasion for feeling the injury. Let us see at this time whether it is more laudable to take revenge by arms or to draw an honourable satisfaction from it. This question is worthy of debate because many, when they are offended,

---

[186] A contemporary card game

do not want to hear of any accord. It is the ordinary way of the noble gentleman that when he is offended in his honour, it is necessary that weapons provide him satisfaction, but also, if his party reclaim for him an honest satisfaction, he must be satisfied with it and that one must be considered as sufficiently content as if blood had been drawn. The quarrel, too, is very often of such importance that there are no arbiters who can reconcile them and they are forced to come to arms. The King must acknowledge it,[187] and, if the honour of one were so greatly offended that an agreement cannot be reached, it is very reasonable for them to end it through arms. But where there is only a small difference and honour is not offended, it must be reconciled by the voices of honest gentlemen, their neighbours. At this time, for the first report that we makes they are at arms and killing themselves for very little reason. I say for yes or no we consider ourselves to be denied and insulted, or braved, and then they say that their honour is offended. It would take a censor[188] to correct the lightness of their quarrels, [those] who expose themselves to death without distinction or judgement. I have proposed at the beginning of the first part the causes and the reasons why the Prince must grant combat. Also, I said that it is not allowed to take this revenge indifferently to the injuries. I believe that many will find this strange, being of bold and valiant courage, not wanting to let go of anything that touches their honour. This consideration has a lot of strength, but also, if we want to heed that murder is never pleasing to God, we will find that there are many good reasons which command us to take no revenge because God has reserved this for Himself. This is the reason why I want to say that this proposition is divine, in which it is necessary that the honest gentleman stops himself, fearing to fall into a peril so great that the vehemence of his courage, and his too furious anger, do not make him stumble into a misfortune so great that his honour, his reputation, his life, and everything, his property and his family cannot but fall into dishonour, and into an infamous reputation, and that this will cause him to be despised and dishonoured by everyone. There is no man of such little judgement who does not know that murder is displeasing to God and that the gentleman of honour must avoid it if he wants to perpetuate his honourable reputation, other than repelling the injury done to us or in the service of his Prince. Socrates said that it was not just to offend someone, even though he had outraged us, especially since the good man should never do evil. And nevertheless there is no

---

[187] *le Roy en doit prendre la cognoissance*

[188] *censeur*. Cotgrave has "A censor; or comptroller, a master of discipline, reformer of manners, punisher of disorders."

man of whatever condition he may be who is not full of this desire for vengeance. This is the reason why a man must be magnanimous to correct this violence and extreme passion and to overcome it gently. Alexander was accustomed to say that it was more necessary for man to be filled with a big and magnanimous heart to forgive his enemy the injury he has done than to kill him, meaning when the man is filled with spite and revenge, that he should not have the desire will to carry out his evil design, but desire to correct it and to temper his anger. This is more difficult for him to do than to undertake to kill a man. In fact, the appetite for revenge is a vehement passion, which boils the natural blood which is in the man, and which causes it to be transported and to be altered to do unreasonable and illegal things. It is the reason why it is very difficult to moderate. Also, the he who pardons those who harm him is worthy of honour. It is not very wise to be moved by the first words that one says inappropriately. One must take care not to always obey our courage, which is often so prompt that, if we do not moderate it, we shall be at all times perturbed and without reason. But I am of this opinion that the injury which is done by force of arms, must be repelled by force of arms, and that he who is assailed must be excused if he kills his enemy while defending himself. Also, when a murder is made by ambush and betrayal or assassination, it is not excusable. But, if the gentleman was the weaker, and alone and hurt by his enemy who was better accompanied, if he avenges himself in the same way, he will not be at fault. Also, when his party uses this reason give him an honourable satisfaction, admitting to have failed and done a deed against all the obligations of a gentleman of honour, I consider that he must consider himself content and well satisfied. We we have spoken about it in the chapter on satisfactions. It is not necessary that one stomach something of small consequence, but one should respond honestly and without passion. Lysander, one day being injured by a wicked man, told him, "Vomit boldly, my friend, and do not spare yourself. By doing so, you will be able to empty your soul of the thousand evils of which it is full," meaning that he did not feel injured by the wicked man. I would therefore conclude that the gentleman must avoid revenge and that it is more honest to derive his reparation by seeking honest satisfaction, and accordingly as the offense and the injury merits.

Chapter X – That shame and dishonour must prevent the gentleman from doing wrong

Cleobule[189] said that the happiest city was one where the inhabitants held dishonour and shame in higher esteem than law. We spoke above of the virtues which the gentleman should follow. It seems to me that it will be very appropriate to say that the gentleman who loves glory and honour must flee from shame and dishonour. Plato affirms this saying when he says that the desire for virtuous things and the shame of dishonest things make the man live happily and with honour. I do not mean that the shame must be such as it repeats[190] the person who is fearful and timid and without any honest grace. But I mean to speak of the one who prevents evil words and evil deeds which bring dishonour to the person, as being contrary to the duty of a good man. Socrates is of opinion that virtuous shame is becoming to youth. So, when it is said of a young man that he is honest and with good grace, it is properly virtuous shame, and that which I want to talk about. The young gentleman should not be fed with such freedom that this license makes him impudent and of bad grace. For the wicked man is never ashamed, and when one wants to correct and condemn his vices, he takes no correction and is not ashamed of them, laughing at all that is said to him. This is testimony of the very bad nature which is in him and which is totally addicted to vice. This is why parents should often condemn the vices inflicted by them and make themselves aware of the perils and shame that can befall them by continuing to do wrong. Parmenides[191] taught his disciples that there was nothing so dreadful in a magnanimous man as dishonour. The Persians had their children educated so well that they would not allow them to do a single dishonest act, and had killed whoever would have put himself naked in front of another. Still today, the honest and virtuous shame is in such strength with honest gentlemen that you will not see the son laud himself with his father or pass the door in front of him. It is honest shame that stops them and corrects them when they are too quick and vehement. Not without reason have I discoursed on this question: for virtuous shame is most often the cause of having

---

[189]Likely Cleobulos of Lindos, fl. 6th century BC, poet and philosopher, one of the Seven Sages of Greece.
[190]*redist*
[191]Parmenides of Elea, fl. late sixth or early fifth century BC, who also appears in Plato's dialogues.

quarrels and, for the fine differences which are in people, we do not tickle each other so often for this quarrel as do those who are successful and who have no shame or judgement in what they do. I do not want, however, so much to recommend shame to the Gentleman that it would come out of him as a vice which was reproachable to him. For shame could be such that he allows himself be crushed and melted into dishonour. This shame should be esteemed as vicious. Those are lacking in magnanimity who allow something that is repugnant to their honour to pass, being accompanied rather by great cowardice than by virtue. This is how shame is praiseworthy when it corrects vice. Also, it is held to be very pernicious, when there is neither prudence nor honour which can correct it.

Chapter XI – Of the fear that accompanies shame

Most often, fear of displeasing the great is the circumstance in which one does not dare to speak the truth of what one knows and to advise on it. Thus, through lacking heart and for fear and shame, one is not bold to profess what one is required to do. This is an opportunity for some to fall into faults which cause them dishonour. Also, when one is called to give advice, one must always aim at this goal of always being true, and not consent to evil. Otherwise, it would be to declare openly the weakness of his heart. The man of honour, bold courage and authority will always speak roundly, and will pronounce without shame that which is right and reasonable, without disguising his words. Plutarch has made a treatise on *False Shame*,[192] and teaches us a very honest doctrine to avoid it. He says, if someone begs you to do something for him and you cannot do it, assure him of not doing it, and take care not to deceive him. And if he communicates some matter to you, which you do not understand, do not be ashamed to tell him frankly that you are not well versed in this subject. And if you are well informed, do not be afraid to tell him your advice freely, without any dissimulation. Flee from all lies and do not deceive your friend with a false opinion. Zeno, this great Philosopher,[193] one day met one of his friend who was walking alone. He asks his friend what he was doing there. "I am here at Lescart," he says, "because one of my friends wants to employ me to do something for him, which I

---

[192] *la Honte vitieuse*, better known under the Latin title, *De vitioso pudore*
[193] Probably Zeno of Citium, the founder of Stoicism

do not want to do, especially since it is an unfair thing and against all right and reason." Zeno answers him, "Are you really so ashamed and so fearful that you cannot refuse him, since it is something that is neither just nor reasonable? Have no shame in refusing him, as he had no shame in asking you." Agesilaus[194] was hassled by his father to give a judgement of a cause which was before him, and rather iniquitous. He answers him, "Father, you have taught me from my youth to obey the laws. It is reasonable that I follow your first lesson and take care not to go against duty and equity. You must not, however, allow yourself to be so carried away and entombed in this fear that it makes you forget the care that affects your reputation." The greatest captains, who learned that one wanted to tyrannise them, for the shame that they had that someone thought of them that they were scared of death, they did not want to remedy it. Caesar was warned by several times that one had conspired for his death, even the day he was killed, nonetheless, he despised him. Enough others having despised him did not get on well.[195] I will say then that the Gentleman must embrace that virtuous shame which will keep him from stumbling into vice. Where there is shame, there is virtue which keeps people from doing wrong and from falling into dishonour, and acquiring a bad reputation.

Chapter XII – Of the fear and false shame which the gentleman must avoid

Formerly, the Romans had their laws so praised that they observed them very rigorously and without having any respect for the good or evil which should result, provided that they were observed. We have an example of this which is very worthy of being noticed. The son of Torcatus,[196] lead an army and had the enemy before him ready to fight, and saw victory in his hand. Yet the Senate ordered him not to fight. However, this did not stop him fighting and obtaining the victory. Torcatus, then Consul, gave his son a triumph in his camp for having obtained the victory. And then he had his head cut off, following the ordinance of war for disobeying the prohibition which had been made to him. This son fought only to avoid the shame and the dishonour that he believed could befall him if he did not fight his

---

[194] Probably Agesilaus, King of Sparta, also recounted by Plutarch in *Parallel Lives*
[195] *Assez d'autres pour l'auoir mesprisé s'en sont mal trouuez.*
[196] Titus Manlius Imperiosus Torquatus

enemy, seeing the victory in his hands, hoping by this means to do a signal service to the Republic, and not for impudence or too reckless inconstancy. In this, however, the father did not abuse his authority or his power but it was military law that compelled him to do this, for his son's fault was in not having obeyed the commandments of the Senate. Captains Epaminondas and Pelopidas were condemned to death for the same reason,[197] knowing the advantage they had over the enemy, and that the Republic was lost if they did not give battle, from which they obtained a signal victory, and by this means preserved all the State of Thebes. Although the people gave them grace, they were nonetheless condemned, the benefit, the utility and the honour of the fatherland was the principal reason for it. There are some who are much worse off, and do not put before their eyes the shame they should have when they undertake to kill and assassinate for money. Shame and dishonour should prevent them from do such a vile act. Thus why fathers should be very desirous not to allow their children vicious things. For when they have achieved their strength, it is no longer time to present them any shame, which is overflowing in their actions, and they are so perverse that their father and all their family receive a lot of trouble from it. And thinking of receiving pleasure and contentment in their old age, someone comes to report that the children have been killed in a private fight, while debating a very small and light quarrel, or for their friends have exposed them without any quarrel. Virtuous shame is properly having fear of wrongdoing and of one's renown falling into dishonour and reproach. It is that which the honest and virtuous Gentleman must embrace. By this means he will avoid quarrels, and by his graceful bearing he will be received and well wanted in all good company, loved and well respected by everyone.

Chapter XIII – That the Gentleman must keep from talking too much because from that comes the increase of quarrels, especially when one speaks inappropriately

It is a very great vice in the Gentleman to talk too much. Also, it is a fine virtue to speak well and to speak well at the right moment. He who knows well how to control his language and speak with discre-

---

[197] Thucydides says these 4th century BC Theban generals both died in battle.

tion has a great gift of Nature in him, not to speak for profit,[198] or with too much license, letting the words flow in an adventure and without judgement. Those who take pleasure in speaking badly, and saying bad things about everyone, are greatly to be despised. In this, we discover in them that there is a lack of good understanding. From this, there comes a great abundance of quarrels, for having offended someone by word, we must change our language and make heavy excuses. This is the reason why the Gentleman must think carefully about what he says, and consider what is good to say or to conceal before he opens his mouth. One must not, however, disdain someone who speaks too often, provided that his language is good and worthy of being listened to by everyone. There are some who are so eloquent that they know how to tell that what they have seen and heard with such gravity and such beautiful language that the whole company can benefit greatly from it. Also, it is required in conversation to take care not to mix in words that are unruly and dishonest, and not to speak ill of anyone. Otherwise, the speaker would acquire the reputation of a slanderer and a great prattler. As one day being in an assembly of honest Gentlemen, who were discussing matters of consequence, one among them, wished to get involved in persuading the whole company and bringing them to his point of view.[199] He was so enamoured of this that he did not permit a single member of the company to speak. One of this company says to him, "Your language is beautiful, and your persuasions are elegant, but there is neither foundation nor reason. Therefore, I am of opinion that we do not follow all these beautiful evidences that you give us." This speech resembles that of Lycosthenes, who endeavoured to persuade the Athenians to go to war with a bold and audacious harangue. But Phocion answered him that these remarks resembled the cypress, which is a beautiful and tall tree, but bears no fruit that is worthwhile. Bias[200] used to say that the tongue is the worst and the best part of man. For if one knows how to lead it well, it serves to admonish and instruct others. Also, if it is bad, it only serves to ruin him. Isocrates[201] said that there were two ways for a man to speak: one when necessity commands him, the other is speaking of that which one knows. For he who speaks often, when his words are instructive[202] and he speaks of lofty things,

---

[198] *aduantageusement*

[199] *persuasions*

[200] Almost certainly Bias of Priene

[201] Isocrates, 436–338 BC, was one of the ten Attic Orators and was very influential in defining the rules of ancient rhetoric.

[202] *sentencieux*

he cannot speak too much because from him one learns a great deal of honour. Alexander the Great gave some money to Cherille, who was a very ignorant poet, so much so that he no longer gets involved in describing or speaking. Silence in time and place, and speaking on purpose are things that are highly commendable. Certainly, I believe that those who know how to speak well, with reason and discreetly, know how to be silent when the time is right. I know some who do not have this gift of Nature, because their affections ordinarily transport them, which is the cause of them not controlling themselves. No passionate man has good reason, especially since the mouth speaks the abundance in the heart, which better discovers our thoughts and the mores of which we are composed than do the traits of our face. Alciat has made an Emblem of a man who has his finger in his mouth,[203] wanting to signify that it is necessary that he is sober in his speech. Also, the words which have been concealed have benefited more than those which have been said and divulged. The histories give us many examples of those who repented for having spoken too much. Sulla took the city of Athens,[204] having been informed where the weakest place in the city was by a spy who had heard it said by those of the city itself. That is why it is a very great wisdom to be able to conceal what should not be said. Plutarch agrees with this statement, and says that kings and those who are nobly nourished must first learn to be silent, then to speak. It is a great virtue to hide a secret well. When Antigonus[205] was questioned by his son, at what time should break camp, answers him, "You will know it when the trumpets sound." By this, he gives this warning to everyone: even though he was his son, he had to learn to be secret in such matters. Cecilius Metelus,[206] was also questioned by one of his Captains, about what he deliberated to do the next day. He answered him, "If he thought that his shirt, which was closest to his heart, could reveal his secret then he would strip it away and have it burned instantly." I greatly regret the levity of many of the nobility who, for the indiscretion of their speech, are often in quarrels without looking at the weight and the merit that may result. The honest way of speaking is therefore to be prized. The grace, countenance, posture, gestures, all these beautiful circumstances are worthy of the honest gentleman. By these beautiful features, one recognises

---

[203] Andrea Alciato, most famous for his work *Emblemata* (1531)
[204] The Roman general Sulla sacked Athens in 83 BC.
[205] Antigonus I Monophthalmus, "One-Eyed", 382–301BC, was a prominent military leader in Alexander's army, he went on to control large parts of Alexander's empire after his death.
[206] Quintus Caecilius Metellus Pius

brilliance in the man, and something of the magnanimity which one notices in him, which is agreeable to everyone. I am of the opinion that to speak inappropriately is the cause of abundance and the multiplication of quarrels. There is as much to rebuke in someone who writes inappropriately as in someone who speaks without consideration, which we shall speak of in the following chapter.

Chapter XIIII – That the gentleman who writes inappropriately is greatly to be condemned, because from this arises many quarrels

There are found many who have been reprimanded and chastised for having written what should have been hidden. The letter written by our hand is nothing but the message and the interpretation of what we want to do, or else it is make known the truth in our absence of what we would like to say if we were present. That is why when a man writes, his letters must be full of wisdom and good advice, as if we were saying it in person. It is a gift of grace to know how to put it into writing, and something everyone should study how to do. Long letters mostly just repeat [themselves]. Brevity and beautiful style are the most excellent. The ancient Romans were accustomed to it. Pompey, writing again to the Senate, sent word to them in these terms, "Peres, Damascus is taken. Palestine vanquished. Pentapoly subjected. Syria, Arabia, and Slavonia remain as allies and good friends." Plato, writing to Denis the Tyrant, sent him in six words the wickedness that he was full of in order that he corrected himself, "You are a parricide for having killed your brother. You torment your people. You impose intolerable taxes on them. You are served by the wicked. You hate good people. You lose all your friends. All these things are the office of a tyrant. Take care of yourself and correct your morals." It was the way of writing to which the ancients were accustomed. However, a too great brevity in writing can only be done with a beautiful style. A certain eloquence and beautiful words are required for it [and] the letter would be a lot more pleasant. And, be careful that in writing it offends no one and that nothing is put in it from which afterwards it should be taken back. In wanting to teach the gentleman to correct himself for talking too much in order to avoid quarrels, I want to warn him to take heed of the letters he writes, so that these letters do not make him stumble into some serious fault for which he would be forced to repent afterwards. Marc-Anthony put Cicero to death for having written many evils and invectives against him. He had his head and both hands cut off because he had written the Philippics,[207] which still today are perfection. Salust,[208] who was a great orator, had this imperfection to write well and badly against everyone. For this reason, he was forbidden to write any more, being a very vicious thing for an author to rail against everyone according to how his own passions dictate. This is why we must take care when we write letters that we don't put things in them that offend anyone. Because, even when we have spoken inappropriately, there is a way to smooth over our language, whereas the letter written and signed by our hand is a testimony of the truth which we cannot renounce.

Chapter XV – That ingratitude is a vice which the noble gentleman must avoid because he who is ungrateful procures many enemies

The words of Sophocles are greatly to be noted when he says that a man must have memory and remember from whom he has received pleasure, because according to his judgement the ungrateful man cannot be considered just or noble. Very often one remembers the evil and the displeasure that one receives but the good and the pleasure is immediately forgotten. This is totally unworthy of the gentleman of honour and virtue. The ungrateful man will never be loved or esteemed by anyone and will be held as brazen and impudent, having no knowledge of the good and the honour he has received from his friend. One who is stained with this vice must be rejected from the company of honest men, like the one who takes no care to recognise his friends or those to whom there is an obligation. Because there are some who have such bad natures who, after having received honour and much pleasure from their friends, are so reckless, forgetting all the honour and the office of the honest gentleman, that they have tried to harm them in everything they could, so perverse is their nature and their courage wicked. Is it not the office of a barbarian to do such acts, similarly the gentleman who should be accomplished in virtue and honour? Necessity sometimes compels most men to seek favour,

---

[207] A series if fourteen speeches written in 44-43 BC condemning Marc-Anthony.
[208] Roman historian and politician of the first century BC.

help and support from his friend and, having received it, should he forget the bad offices done him before now? Today, however, the gentleman is accustomed to it, when one wants to correct him for this fault, he has his reasons quite ready, which he parades and wants them to be taken for legitimate. But if someone has caused him displeasure, he wants to avenge himself. The good deed is immediately put under the feet, so ungrateful is he. It is quite certain that he makes many enemies and because of this finds himself abandoned by friends. On the contrary, he who is not ungrateful, because he recognises the good and the pleasure he receives from his friends, is assured of having his mind at rest, and being well esteemed by all, so that he can be assured of being helped in his affairs. We must not, for the slightest displeasure that is caused to us, forget the pleasure that our friend has given us, making the son complain of his father, the brother of his brother, the friend of his friend, the servant of his master. The great are even affected by this vice of ingratitude. Draco,[209] in all his laws which he ordained for the Athenians, made a law against the ungrateful, in which he said that if there be found anyone having received a benefit and for which it was proven that he was ungrateful, that he was to be put to death. Ingratitude was so hated by people of virtue that Alexander the Great, who was extremely liberal, never gave a present to an ingrate, and Caesar, who willingly forgave insults, did not give grace to any who were ungrateful. This vice is so detestable that it engenders quarrels. The ancients had this odious vice that they fought over who was of more honourable service to their companions, and even to their enemies. I will speak of one who is worthy of being produced as evidence above all the others for doing a very magnanimous and very memorable act. Cicercius, who had been the secretary of Scipio the Great, saw to the awarding the status of Praetor to the son of Scipio by the common voice of all the people and of the Senate, recognizing the honour he had had from his father. He, not wanting to be ungrateful, desisted from his pursuit and solicited for Scipio and having won him this status, by this he acquired the reputation of being very virtuous, for having wanted to serve the son of the one to whom he was obligated and not having shown himself ungrateful for the good and the honour that he had received from his father. This is an act certainly very generous and which must be recommended to all those who will want to be ungrateful towards their friends.

---

[209]Draco or Drakon, 6th century BC, replaced the oral and traditional law of Athens with a written constitution enforceable in the courts.

Chapter XVI – That the Gentleman must not reproach his friend for the pleasure he has given him

The generous Gentleman, although he knows that people are ungrateful for the good turns he has given his friend, must not stop pleasing those who want to employ him, and must be careful not to use reproaches, especially as a reproached [or tainted] service can never have good grace, and are half sold. There are, however, those who are of such a perverse nature and so ungrateful that one cannot be commanded other than that one use any manner of reproach. That which must be said is a retelling of the good deed that one has done for his friend who is ungrateful. To these, he must make them remember the service that has been given them, not in order to reproach them, but to make them ashamed and disparage the ingratitude with which they are full. The ungrateful man is nothing but the one who tries to get from everyone all the profit and convenience he can and loves no one, not wanting to recognise the good and the benefit he has received of his friend. Is there any greater ingratitude than that of Phocas,[210] who during the lifetime of the Emperor Maurice was only a centurion, being possessed of this vice caused his master, his wife and his children to die and made himself emperor. Forgetting all the honest duty that a loyal servant was bound to do towards his master, and ungrateful of the good and the honour he had received from him, he dispossessed Maurice of his empire to seize it. Here is a traitor and a very ungrateful servant. Republics have been greatly condemned for showing themselves ungrateful to those who loved the welfare of their homeland. Cicero, that great orator who had done service to the republic and prevented the conspiracy that Catiline had conspired against the Senate, was once exiled from Rome for the troubles that one brought to him, and later recalled for the regret that the people had at his exile. Solon, this great legislator of Athens, was driven out without ever being able to return there. Publius Lentulus, who had virtuously defended the republic, was banished from Rome and on his departure prayed to the gods to favour him such that he could never return to a people so ungrateful. Enough has been said about ingratitude, and the harm that this vice can bring, in order that the honest gentleman will take pains to avoid it. In doing so, he will find himself at rest and his mind full of ease and contentment. By this means, he

---

[210]Leader of a military revolt which overthrew the Byzantine emperor Maurice and established himself on the throne in AD 602.

will acquire many friends. On the contrary, if he is ungrateful, he will get many enemies.

Chapter XVII – That the poverty of the Gentleman should not cause him to be poorly behaved, so that he does not fall into disgrace

I know well that many hold poverty to be an evil very difficult to bear, and that there is no disease so grievous as poverty. I am of this opinion that the most cruel enemy that the Gentleman can have is poverty, and that this necessity makes many desire the goods of others, such that many have an insatiable spirit. But when this disease is recognised and held with good reason, I find that it is easy to cure.[211] And that for this reason, the Gentleman must not form a habit of evil doing, nor of doing acts which are not the duty of the good man, and making himself bad mannered. Because otherwise, he would be badly spoken of and would acquire a very bad reputation. Many allow themselves to perform bad deeds, and afterwards they fall into very dangerous circumstances, and all result from want of embracing virtue, which ordinarily prevents the Gentleman from stumbling, and serves as a bridle for him, restraining him when he wants to do some act bad and unworthy of his profession. Also, if he does not want to contemplate this beautiful virtue and be satisfied with the means that God has given him, he will find enough meat and matter[212] in order to debaucher himself, and to do things illicit and against all the duties of an honest Gentleman, so much does he have a haughty and unruly heart. There are many who do not care how they should obtain goods and, without taking care of their honour, they govern themselves as they please provided they get them. It seems to me that when the Gentleman wants to measure himself according to his means, he will be able to live honestly, provided that he does not put himself at excessive expense. And that one can say of him, that he is a man of substance and honour, and who lives with an honest reputation. There are so many contentments in a small household that many desire it, so as not to have their minds agitated by so many troubles as the rich most often have. The Gentleman who does not have great means

---

[211] Original text has *entretenir*: maintain, support, etc
[212] *assez de matiere et de subject*

can exempt himself from legal process if he wants, and should avoid them as long as it is possible for him and, also, avoid a great superfluity of expense and clothing. Is this no more honourable, and approaching more closely to virtue, than to seek baser conveniences, which could in the end only serve to dishonour and ruin a poor house? It is something that has always been practised among gentlemen of little means, to put their children in the service of princes and great lords, as I have said above. This custom should not be lost, and I would always advise him not to disdain to help, and not to look so much to his nobility but only to the necessities of his house, bringing benefits to it through his work. Also, the benefits which are acquired through good conduct, the achievers possess them for a longer time, the house prospers further, and even the family line is better esteemed. Having ill-gotten gains, the family never prospers from them. We have endless examples of those who are satisfied with little, and who have better valued poverty than wealth. Archelaus was in the habit of saying that however much one's own poverty seems to be annoying and hard to bear, so it is nonetheless the true school of all virtues, and produces good and virtuous families. Euripides said that the rich were full of vices and the poor full of wisdom, because poverty makes a man more nimble and his mind more agile, and men greater and more excellent at all that are proper and necessary to the lives of humans. So said a Philosopher, that it was not necessary to flee from poverty, but especially from injustice, for the poor man, when he is just, is full of virtue. On the contrary, if he is full of injustice, he is a monster full of wickedness. This is why it is more necessary to live with few goods in peace and patience than to have many, and to enter into continual torment and anger. Epaminondas and Lycurgus were not esteemed for their wealth but they were honoured for being poor, and having greatly benefited their country. To prove that the poverty in some is highly commendable, I will bring back here a response that Diogenes made to Alexander, on being visited by him. And, after much talk, he said to him, "Diogenes, ask me what you want and I will give it to you, because I know that you are poor." To which he replied, "Alexander, which of us two do you think is poorer: me, who is content with what the gods have given me, which is a small thing, or you, however much you are King of Macedonia and, not content with that, you want to extend your domination to the confines of the whole world, so great is your ambition and desire to reign." Then this great monarch, out of admiration, said, "If I weren't Alexander, I would like to be Diogenes." Also, when a friend wants to wish something good and excellent for his friend, he must not wish him great wealth but only that he may have good health and be maintained in honour. Also, he should wish that he take care that he not fall too far from it [into] great need, especially since he who is adorned with beautiful perfections is rich enough and should not want to wish for more, because wisdom and justice will make him prosper so much that whatever poverty he is in, he will nevertheless never lack for benefits. Aristides, governor of the Athenians, usually said that it was only those who were poor in spite of themselves who should be ashamed of being so, and that it was more laudable and greatly to be prized to bear poverty virtuously and magnanimously than knowing how to make good use of wealth. Also, poverty should never be accompanied by shame, except in those who have had many goods in their hands and have not been able to govern them well in such a way that they have fallen into need. I know of those who, after having made much and spent much, died in poverty and in debt, having dissipated all their wealth. In these, poverty cannot be commended. But, in he who is foreseeing, it must be esteemed virtuous, as he who gives proof of wisdom, who does not want to bury himself in vile and shameful things, and who thinks he is able to lessen a single mark of his honour such that his heart of full of courage and his spirit filled with great things. This is why I have been desirous of instructing the Gentleman, for fear that poverty therefore cause him to be badly conditioned, to live according to his quality and not make such great expenses which exceeds his revenue. And that then, after doing otherwise, he was compelled to do harm, and by this means to acquire a bad reputation. We will talk about it in more fully in the following chapter.

Chapter XVIII – That the Gentleman should not set his heart on riches, if not to use them and follow virtue according to his quality[213]

The Ancients and all those who have acquired the reputation of being magnanimous have held virtue in such esteem that they never allowed it to be stained with any vice, nor with a blemish that could lessen it. The great personages, who by virtue acquired the title of virtuous, on this happening, they fled this vice as far as it was possible for them. Having this subject in hand, I have been very desirous to instruct the Gentleman in all honest conditions and ways of living well, that it may

---

[213] Incorrectly labelled Chapter 17 in the original.

make him perfect and well educated, and that he may be praised by everyone, both of those who are lovers of virtue, and of those who have acquired this laudable perfection. I am only focussed on the riches that a Gentleman often sets out to acquire, considering himself miserable when he finds himself lacking possessions, only wanting to parade wealth, most often forgetting that on which the honour and duty of the virtuous Gentleman depend. In the end, we would like to diminish the honour, the virtue, and the reputation of a Gentleman, if he is not accompanied by riches. This is something that is practised enough today, and with difficulty one puts a price on men of virtue, so much does wealth blind people. Socrates makes a comparison, as the horse cannot be without a bridle, so the rich cannot guide themselves without reason. Because wealth brings pride to those who possess it, and an extreme envy in amassing it and total lust to know how to keep them, and with great license and excess to depend on them. Diogenes was also of this opinion, that virtue could not dwell in a city nor in a rich house. For this reason, one has great respect and honour for the rich and the riches of a Republic. He believes that virtue and people of wealth and honour will be less respected. Also, Republics are better preserved by virtue and by those who are wise and highly esteemed. They perform greater deeds. However, I do not want to exclude the Gentleman from wealth and make him poor and miserable. But, I want him to have some wealth so that he, and the house from which he has come, can conduct himself and live according to his quality. This makes me conclude and say that riches help virtue a great deal, but there is a way to know how to use them to help, as I have declared above. For the Gentleman, therefore, noble and of brave courage, it is an unbearable affliction for him to be seen that, for lack of possessions, he does not have the means to be able to improve and grow according to his desire. Also, he has virtue in such a recommendation that he does not want to alienate or engage his conscience for amassing vicious means. Thus, an extreme combat and with great weight in the heart, this noble Gentleman, who balances virtue and vice, acquires honour and reputation. I will therefore say that if the Gentleman procures goods and riches, knowing how to spend them well and help himself where his honour commands him, this should not be reputed as a vice in him. But, if he were to amass them for himself alone[214] and through a great and overburdened avarice, this must turn him into great contempt and dishonour. Also, when I say that wealth aids virtue, I mean that it is necessary to moderate this greed and not to

---

[214] *pour son particulier*

be too active in desiring it, other than to use it well. For he who does not put limits to his avarice and his greedy appetite is always poor, suffering and needy. On this subject, Plato says that long life of the avaricious is not joyful, but taking care to amass wealth moderately makes us live longer and with much contentment, for excessive greed gnaws and eats the hearts of those who desire too much. But where is moderation? This embellishes our life and the renown of those who acquire it, because by virtue one acquires moderation, who returns the honour of their house and the profit and utility of it to their children. There are enough houses of gentlemen in this Kingdom who have advanced themselves by virtue and acquired moderate possessions. Others, who have been so insatiable that they have not been able to put limits to their greed are declining everyday. Also, all those who have run their fortunes well and have aspired to virtue have been modest in their greed, and have governed themselves according to the opportunity which has presented itself, without aspiring to anything more than is reasonable, fearing to embrace too much. I could argue the house of the Marquis de Marignan in the Duchy of Milan, a brave gentleman and good soldier to whom the Emperor Charles V gave the town of Marignan for the hope he had of drawing good services from him, in former times was called "Medequi." He was held to be a valiant Captain and well renowned and won battles in the service of his master. His brother was Cardinal and then Pope. Here is a fine example for those who know how to manage their fortune with moderated reason. It is a saying that I have heard said and told by those who are of his nation. I think it would not be out of place to recount it in this place. To return to my first point, there must be no doubt that the lack of wealth distances the valiant gentleman from his good fortune and prevents him from being able to aspire to great things. But also, when he wants to strive to do honourable things, there must also be no doubt that his virtuous acts would make him succeed and rise to great honours. Ancient and virtuous personages held wealth in such little esteem that they despised it completely and never had any other care than to acquire honour and a good reputation. It is written of Marcus Curio, the Roman consul who had obtained several wonderful victories for the good of the Roman republic, and of the triumphs and honours that he obtained from the fatherland, preferring honour to riches, he only had a small farm very badly built where he stayed and for wealth he did not fail to be highly esteemed, believing that it was more to his glory to command those who had a lot of wealth than to have it himself. Alexander, after the victory he obtained against Darius, took all his goods. He wanted to give them to Zenocrates who refused them. Also, Plato says, when he leaves the republic and as he wants it to be led, that he does not want the Princes, governors, gendarmes and soldiers to have any handling of money, but that it is necessary that they be maintained from the common [purse], to better strengthen what he had just deduced. It will not be irrelevant to recite in this place what I have heard said to the Marshal of Montluc[215] at the headquarters of Thionville speaking to Monsieur de Guyse, at the commencement of his youth, his father would have left him his house totally in debt since he died rich in goods and honour, and has left a very great memory of him, and was held as one of the greatest captains of France. Here is how this great captain should have acquired wealth and honour by virtue. In this is an example that the desire for wealth makes the gentleman so well and so dexterously seek his fortune that it makes him magnanimous and well esteemed. I have put in this chapter among a number of others because it seems to me to be proper for instructing gentlemen to avoid quarrels. I thought that it would be expedient to teach him to not put his heart so into riches that it was the occasion that he did not seek from point to point all that on which his honour depends. Wealth and riches are transitory and fade immediately. But honour and virtue remain eternal and are felt through posterity, which makes a house famous and illustrious for a long time.

Chapter XIX – That the gentleman should follow wastefulness

We said in the previous chapter that the gentleman should not set his heart on riches, if not to make use of them with honour and according to his quality. He must at this time be instructed to flee from wasteful expense and, when he has amassed goods, he should not spend them with such great license that he stands poor at the end of his days, and his children after him. For many have amassed wealth that their children have spent very shamefully and without discretion: debauchery, the extravagant life, banquets, gifts, superfluous presents, sumptuousness clothing, gambling, and all. Arrogance is the results of a wastrel who exceeds his means and the means of the gentleman. If he does not regulate it and if he wants to continue it, let him be quite certain that in a few years he will see the consumption of all his property

---

[215] Blaise de Lasseran-Massencôme, seigneur de Monluc, (c. 1500—1577), marshal of France from 1574, known for his military skill and for his *Commentaires*, containing his reflections on the art of war

and, after that, it will be completely dissipated. If he wants to make a good steward of it and settle his spending, I fear that he will not have the time, and that he will have to remain poor and regret his miserable life. I will bring back to this place examples of those who have squandered all their wealth and have become poor, and others to whom it was necessary to give guardians against the bad life that they exercised in their households. To those, the law compares them to the furious and those who are insane. To these, the management of their wealth should be forbidden. The son of Fabius Maximus was deprived of his father's wealth, because he spent all that his father had left him on banquets and luxury. Because this family of Fabians was a noble lineage and well renowned, the parents were grieved to see its fall from splendour and dignity. Solon ordered a law that those who would be found prodigal and who would have dissipated all their goods were to be held infamous. Also, wastefulness is nothing other than that which has no beginning nor end in its spending, nor any respect in its sumpuosities, and which without reason eats, decays and wastes all its wealth, and for its pleasure indulges in all these sensualities. Cornellius Lentullus[216] who was of noble family spent all his property foolishly and a large sum of money from the public. I know of some, however, in this Kingdom who have spent a great deal of their means and have made houses poor, and who have not been lavish and have not consumed their wealth in gifts or in presents or in gambling, but have run at Fortune (if one wants to attribute it something) and to make themselves known and noticed by the notable acts to which they had aspired but could not achieve. This could not be done without great expense. Those should not place themselves in the rank of wastrels but, indeed, in the rank of those who wanted to try to attain honourable perfections, to which it is customary for the virtuous gentleman to seek in order to acquire honour. It is this Fortune which was not favorable to him. And in remaining poor, he should not be put in the rank of wastrels nor of the infamous but in the rank of Chevaliers of Honour. Having been found in this opinion, I ask which is the least vicious, the wastrel or avaricious? For my opinion, I say that the wastrel is very much less, because by him many profit, because he gives his property freely and with a manifest will to want to please everyone. But the avaricious is only acting for himself and does not exceed his means any more than if he had none. That makes him very ugly, because one sees from day to day avarice grow in him, where the wastrel, when he sees his wealth diminish, looks at

---

[216] Publius Cornelius Lentulus Spinther

the means of recovering the loss and spending he has done, whereas the avaricious is always infected as if from an incurable disease.

Chapter XX – Of Recklessness

Isocrates said that strength with prudence profits. But also, if strength is not accompanied by prudence, it is vicious, because to hazard one's strength without purpose and without just cause is pure folly and foolhardiness. The opinion of Aristotle is that he who says or does something thoughtlessly should not be considered wise but reckless. Foolhardiness also is nothing other than to start and rush in for one's pleasure, without any regard to the danger and peril that one knows to be evident, and to undertake all things without consideration, and to attempt what one knows to be certainly perilous. The virtuous gentleman and the one who aspires to imitate him must consider well being moderate in all his actions, in order that the too hasty movement of his desires cause him to fall into shame or reproach, or into some another big mistake. I have spoken above of boldness, and of that with which I would have the gentleman accompanied, and have spoken a little of temerity, because I could not enrich boldness in its virtue if I did not blame the vice which corrupts it with its opposite, which is temerity. But it was with few words. At this time I will speak more appropriately so that the gentleman of honour, who wants to accompany himself with this beautiful virtue of magnanimity, may be informed of what is necessary for him to do to live honourably and happily. That done, I think I have given him enough fine and good instructions to avoid quarrels. I know that first inclination are not within our power and that we cannot easily correct them. But also, when we know that they harms us, and that our promptness most often causes us to fall into misfortune and forces us to do things unworthy of an honest gentleman, this should give us a means to correct ourselves. Otherwise, it would be characteristic of brute beasts, not able to recover from a fault or similar thing in the future. For to remain always in the same mood, and to continue in it, one must not call these persons men because in them there is no reason. But certainly those who behave thus in their passions, and only want to be subjected to their own will, we must consider that they are filled with too great glory, only wanting to believe their advice and remain firm in their opinion, so that it is not permissible for anyone to be able to divert them. Such people need no advice, so great are their hearts. Also, it is quite certain that to undertake something so hazardous, there is neither praise nor esteem. Cato the Wise replied to a friend of his, who greatly praised a character who was too adventurous and too bold and without discretion, that there would be much to be said of one who had a lot of virtue, and of one who did not trumpet life, as if he had wanted to say that to live and die with honour and virtue is a highly praiseworthy thing and that to avoid death without being accused of cowardice one must not be reproached. Recklessness, strictly speaking, comes from the nature of man. There are some who are the spirit of vainglory, who see themselves much prized, forcing such a high opinion of themselves that they esteem themselves more than they should. It is the occasion that they do and undertake more recklessly what is contrary to their duty. These flatterers, who often go to them praising their actions, and who esteem them for more than they are, are fit for them. In these, virtue does not dominate, because they allow themselves to be carried away by the opinions of others. This is why the gentleman must not lightly believe these flatterers and, because he is disposed to take up arms and aspire to acquire this title of captain, must take care not to fall into this vice of impetuosity, in that there is nothing so certain nor anything that is more to be condemned in a commander of an army than to be overcome by this vice, which causes so much destruction to kingdoms and armies. It is sufficiently read in the histories that several captains who recklessly undertook combats achieved little effect. Some lost their lives and honour there. Others were routed.[217] I want to conclude by this: that those who are called to counsel the great are not so daring as to give them advice which is not useful and worthy of their greatness. And when they realize that the great will be too agitated by passions, they must with great persuasion warn them not to execute their evil designs, so that the Monarchy may be governed by wise counsellors and not reckless advisers. Otherwise, it would only be a deplorable confusion. Aristotle says that man must work to build noble enterprises, being accompanied by boldness and greatness of courage, with good practical knowledge,[218] and, besides all that, with fine industry and patience, remaining firm in his plans, and with good reason and great consideration. Otherwise, those who do not follow this path deserve to be called barbarians and mercenaries of Princes, and will only serve to destroy and ruin a Monarchy, and to present a thousand bad conceits[219] to a noble Prince. And in an army they are advised to deal with[220] the strength of the enemy on

---

[217] *les autres ont esté mis à vau-de-route*
[218] *experience*
[219] *à bailler mille mauuaises inuentions*
[220] *pratiquer*

the day of a battle. It is well known that treason will not take place in a magnanimous and virtuous heart. I will never praise or hold in any esteem the captain or chief of an army who deals with the forces of his enemy for the day of a fight. In this, there are no cunning (although some approve of it). For if one must be praised and esteemed for subtleties and treachery, one must no longer put a price on the honour, valour, boldness and fine conduct that great Captains insisted on observing in the conduct of armies. For the greatest coward, having the title of Captain and having never exercised arms, in this way will obtain victories. In this regard, I will cite again the magnanimity of Camillus, who was dictator in Rome, holding the city of the Falernians besieged. The Governor[221] of the children of this city, having the most riches in his government, and pretending to want to take them outside the city, wanted them release and give [them as] prisoners to Camillus. But this virtuous chief made a very worthy response, saying, "Although in war we use many evils and outrages if necessary, between those who are magnanimous and of virtuous courage keeping honest reason is preferred, governing themselves with justice and the honour of the war. A great Captain should make war trusting more in his own virtue rather than in wickedness of others. It is not necessary to be so desirous of a victory to obtain it with such mean and infamous means." By this this great Captain had more esteemed justice, his honour, and reason more than victory. This is how recklessness often makes great Captains stumble when they are possessed by it. The Gentleman who wants to acquire a name as a magnanimous and excellent Captain must avoid it. And even if he cannot aspire to such a great rank, [he should] flee quarrels with companions, and keep to his friends. Regardless, he must moderate his actions so that he does not fall into this vice of recklessness.

Chapter XXI – That memory is excellent to the Gentleman who wishes to follow arms. And that there have been great Captains who have been much esteemed.

Cicero, who is the father of eloquence, says that memory is the divinity of man and the immortality of the soul. Similarly, Pliny and Plutarch say the same and similar to divinity. Memory is the cabinet

---

[221] *Precepteur*

and receptacle of all that we learn, see and hear, and must be often exercised so that use and exercise can strengthen it further. Pliny speaks of the memory of Caesar who dictated a letter, and read in some book, and heard another speak at the same time. He customarily described all that happened under his charge and guidance in his *Commentaries*, which remain to us as a memory. And by this instruction, many in our time have followed this trail of leaving a memory of the facts of where they were employed. You have the Lord of Bellay, who wrote a book where the precepts of war are taught in imitation of King Pyrrhus, who took great pains in composing the teachings for armies and how they should be conducted. You still have in our time Monsieur, Constable Anne de Montmorency, how much he has not put his deeds in writing, and that they are well enough written about in the Annals of France. However, it is there that in his houses all the ruins and seizing of cities that he did are described[222] to leave this memory of him to his house and to all his descendants. And he died with arms in hand from a wound he received at the battle of Saint Denis, over seventy six years old, which is greatly to be prized and made for him in great esteem. There have since appeared those who have done the like. Monsieur de Desse, who was commissioned by the late King Francis I and by King Henry II, had all his deeds and acts traced and described in his house at Espauvilliers,[223] all his deeds and acts which he performed during the war while he had command. It is a memory which shines in all his descendants. And he died, arms in hand, at Thérouanne[224] by a musketeer. I would put in this place that brave warrior, Monsieur, Marshall de Montluc, who left such a memory of himself through his comments that anyone who makes a profession of arms should read it and will find there fine instruction which will inspire him to do similar exercises and to undertake battles, brave and noble. And in order to witness his actions and his high undertakings, he calls Captains from his time, who are a manifest proof that he did not want to write about what he did in his office, which may only be recognised by those who saw him. Also his comportment was executed valiantly and is worthy of memory. He has always been held to be one of the valiant and bold Captains of our time and the most adventurous. Also he died in his bed aged eighty, after having received in all these wars seven fusillades,[225]

---

[222] *sont peintes*
[223] Possibly André de Montalembert (b. 1483), lord of d'Esse, Espanvilliers and la Rivière who was a Chevalier of the Order of the King and commanded the army.
[224] Thérouanne is in the Pas-de-Calais region
[225] *arquebuzades* is the exact term used.

the last being at Rabastens[226] that he took by assault. It is a testimony that he did not spare his life for the service of his master. After having spoken of the memory of the brave warriors, I will speak of the wise and learned. I have seen in my youth a learned man named Romillius, who was blind by nature, teach publicly in Paris at all hours that he was called to do it. It is maintained that Homer and Democritus had their eyes gouged out to have a better memory. This is how memory has been greatly praised in those who wanted to practice it. It is a great praise to see a gentleman talk well about what he has seen and heard when he is in good company. As we have said above in another place, he is much better understood and much better esteemed by it, and his reputation is published everywhere. Such personages are willingly sought after by Kings to command in an army or to be sent as embassies for the affairs of the Kingdom. Also, it should not be necessary to call them to business except those that the king knows to be valiant and experienced, and will be found well served by them who will return to the profit and honour of his State.

Chapter XXII – That the Gentleman in his affairs must seek the advice and counsel of others

It is a misfortune that accompanies humans that they are often quick to give advice and lead others, and cannot take it for themselves. I believe that this fault comes from not being masters of our first inclination, which causes us to fail in our duty. In as much as ordinarily we allow ourselves to be carried away by our passions, we believe within ourselves that there is no one like us. Here is the imperfection of our life. To correct it, it is much more certain to confer about our affairs with someone who is a very skilled, so as not to vary and to be more resolute in them, certainly believing that two people will solve more cases and happenings than one alone. This is something that is known without having made a long trial of it. There are those who, in other's business, speak very discreetly and with good advice and, in their own affairs, they are very ignorant and cannot put them into any order. Caesar, who by his bravery, led so many fine armies and won so many battles and arrived at the perfection of his greatness. However, he could never escape the conspiracy which had been made against

---

[226] On 23 July 1570, the army under Blaise de Montluc took the town and massacred most of the Protestant defenders.

him by Brutus and Cassius, even though he was warned of it. Yet this cost him his life. His successor Octavius, however great an Emperor and very happy he was, was warned of the lasciviousness of his daughters. He could not, however, put them in order. The Emperor Anthonius the Debonnaire had for a wife Faustina,[227] the most excessive in her lifestyle that he could find, and whatever remonstrance he could do, she had to exercise her lust publicly with the gladiator she had loved uncontrollably. Nicias, who was a valiant Captain, never made a mistake with the advice of others, and when he wanted to be served by his own advice for himself, succeeded in nothing useful. Aristarchus said, "Being inconstant, we do not know what to desire, nor what to flee. Truly, it must be concluded that the valiant man who has no other influence in his affairs than his own judgement and his own advice is the ruin of his house and of the Republic." Let the wise gentleman take good care in his actions and not conduct himself inconsiderately and without advice. Otherwise, he will find himself well removed from his judgement. Also, his business will prosper better when he proceeds with advice and good counsel.

Chapter XXIII – That the Gentleman should not be curious about other people's affairs

Plato, writing to Denis the Tyrant,[228] directed him that he who want to know the business of others is more fond of his enemies than of himself. For as soon as he knows of them, he is quick to disclose them and gainsay them. And, in fact, the inconstancy of man is such that he will never have his mind at rest if he does not know the affairs of his neighbours. Not to bring some remedy to them, but rather to make fun of them, if one knows anything about them. It is our spirit that is not idle and has such unrest that it wants to be a participant in all the affairs which happen in the area, even wanting to embrace all things, if it is possible, in order to rehearse them in his imagination. And very often this desire brings with it several discontents who don't want to be put into the common language of everyone. Pindar was asked to know what was most difficult to do. He answered that there

---

[227] There appears to be some confusion here. Anna Galeria Faustina the Younger was the wife of Emperor Marcus Aurelius. Faustina was the youngest child of Emperor Antoninus Pius (The Debonnaire?) and Empress Faustina the Elder.

[228] Dionysius I or Dionysius the Elder (c.432–367 BC)

was nothing so easy to do as to rebuke another. And, on the contrary, nothing more difficult than to be rebuked. Also, there are many who would not want to inquire, nor would they want to know, the affairs of their neighbours, and even though they were told something about it, they forget it, because it is something that does not concern them in any way. Pliny tells that Marcus Porcius was highly esteemed because he never inquired about the news of Rome, nor about what was said and done there. The Athenians established a law in their Republic that no one should inquire, when someone came to their city, to know what he had come there to do, where he came from and what he was asking, under pain of a large fine and of being banished from the country. They made this law to make it known that to be curious to know the affairs of others was so great a vice that one should avoid it. This is why I will say that the gentleman is very happy who does not wrap himself up in other people's business, and who is not curious to know it, because he who mixes the business of others with his own never has his mind at rest. It is impossible that he will not be driven by the sin of envy, or that he will not be held for a scoundrel or a mocker, and in the end will be shown the duty from which one will keep oneself as a man who is a plague in his country. There are those who exercise their curiosity in other ways, and want to know all sorts of discoveries,[229] which are made in all professions. I know some who have applied their understanding wanting to be Alchemists and to make quintessence,[230] even to make the Philosopher's Stone, and have consumed much of their wealth in this. In the end, they make themselves poor and sickly from it, so great and insatiable is their curiosity. The gentleman must flee these discoveries and must apply his understanding to more honourable ends. It is curiosity that desires more.

Chapter XXIIII – That the gentleman must guard the honour of the Ladies and fight for his mistress

Ladies have the privilege of being respected and honoured, and those who turn away from this honest way of doing things, do not observe the rights that belong to ladies. The best grace that a gentleman could

---

[229] *toutes fortes d'inuentions*

[230] *Quintessencer, quintessence*: Extracting or refining the natural energy or virtue from a thing; the spirit or virtue extracted.

know is knowing how to choose a mistress, and he is not considered a valiant man if he is not well loved and esteemed by the ladies. This exercise is worthy of the gentleman of honour and the valiant man. So, when it is accompanied by its beautiful perfections, he is best suited to good companions and much more prized, and must be careful in all things to be careful to take the ladies' side and, by saying bad things about them, he will acquire a bad reputation. All the great captains had mistresses and held them in high esteem. I will recount for example what Plutarch wrote about Pompey, who had a friend named Flora. One of Pompey's friends fell in love with her and his friend asked if he could sleep with her, which Pompey granted him. However, Flora did not want to consent. Nevertheless, through persistence and fearing that her friend might think it was a deception, she agreed. Afterwards, Pompey did not want to see or love her, and she died of regret. It is written that Alexander did not want to touch his friend's friend, even though he was very much in love with her. One must love his mistress and not dishonour her but preserve her honour at the point of his sword. That is is the duty of the valiant gentleman. Julius Caesar had the head of one of his captains cut off for having raped the friend of one of his companions. The rape of Helen, whom Paris kidnapped, was the cause of the destruction of all. Cleopatra killed herself when she saw the death of Marc-Anthony, fearing to fall into the hands of Auguste Caesar's soldiers. It is therefore necessary to return to the examples of the pagans who had so much of their honour in recommendation that they would not permit themselves to be dishonoured or suffer the dishonour of their mistresses. I will therefore say that when the mistress of a gentleman has been offended that he must resent it and avenge the wrong that someone did to her.

LongEdge Press

LongEdge Press publishes quality translations of French texts of interest to the scholar and practitioner of historical fencing. Visit LongEdge Press at www.longedgepress.com for more books, articles on items of historical interest and practical guides.

Secrets of the Sword Alone Henri de Sainct-Didier (1573)

Fencing Through the Ages Adolphe Corthey (1898), including his report on the *Transformation de l'Épée de Combat* (1894) and several contemporary newspaper reports of his public demonstrations of historical fencing.

La Canne Royale Larribeau, Humé, comprising Larribeau's *Nouvelle Théorie du Jeu de la Canne* (1856) and Humé's *Traité et Théorie de Canne Royale* (1862)

The Art of Fencing Jean de Brye (1721)

Fencing Manual Ministry of War (1877)

Manual of Contre-Pointe Fencing Joseph Tinguely (1856) with a foreword by Julian Garry of *De Taille et d'Estoc*

Archives of the Masters of Arms of Paris Henry Daressy (1888)

www.ingramcontent.com/pod-product-compliance
Lightning Source LLC
Chambersburg PA
CBHW030234170426
43201CB00006B/217